THE MANAGEMENT
DEVELOPMENT IN

The Management of Urban Development in Zambia

EMMANUEL MUTALE

Routledge
Taylor & Francis Group

LONDON AND NEW YORK

First published 2004 by Ashgate Publishing

Reissued 2018 by Routledge
2 Park Square, Milton Park, Abingdon, Oxon OX14 4RN
711 Third Avenue, New York, NY 10017, USA

Routledge is an imprint of the Taylor & Francis Group, an informa business

First issued in paperback 2018

A Library of Congress record exists under LC control number: 2003058368

Notice:
Product or corporate names may be trademarks or registered trademarks, and are used only for identification and explanation without intent to infringe.

Publisher's Note
The publisher has gone to great lengths to ensure the quality of this reprint but points out that some imperfections in the original copies may be apparent.

Disclaimer
The publisher has made every effort to trace copyright holders and welcomes correspondence from those they have been unable to contact.

ISBN 13: 978-0-815-39797-7 (hbk)
ISBN 13: 978-1-138-62110-7 (pbk)
ISBN 13: 978-1-351-14604-3 (ebk)

Contents

List of Figures

List of Tables

Preface

I am a second generation Zambian urbanite and have always been proud of being one. My abiding recollection, albeit somewhat vague, of Zambia's campaign for independence is when, as an eight year old, my friends and I would line the streets of Lusaka's Chilenje African township barefoot and in a pair of threadbare khaki shorts and shirt. Party functionaries would from time to time pass through our street chanting: 'upper roll - vote Mudenda; lower roll - vote Mudenda'. This was a campaign slogan which revealed a socially divided society. Not only was the electoral roll arranged according to social class, the town too revealed deep divisions - with European settlements segregated from African settlements. These and more are the politics and economics of Zambia's urban development process which this book inquires into. The purpose is to reach an understanding and contribute to the explanation and management of the urbanization process in Zambia. This book however, was written at a distance – in London - the distance hopefully ensuring objectivity in what could easily be an emotionally embroiling study.

The first part of the book explores the global theoretical debate on the emergence of urban management as an academic discipline and reviews a selected literature on the study of colonial cities, including recent research on urban management issues. This part also traces the development of urban policy in Zambia against the backdrop, successively, of mining company capitalism, colonial and federal government hegemony, a post-colonial socialist ideological hegemony and, since 1990, a renewed encounter with the political pluralism and liberal market policies. The second part is a case study of Nkana-Kitwe, one of the towns on Zambia's Copperbelt, and specifically examines the power structures in Nkana-Kitwe's urban development, exploring aspects of conflict and co-operation between different interest groups and the relationship this has on the distribution of benefits and burdens related to land, housing and services.

The focus on urban policy necessitated the use of a structuralist methodology which relates to institutions for urban management. Because urban policy is a human formulation and also because of the human presence in urban institutions, this gives the research a definite praxiological slant. I therefore eschew the treatment of people as if they were inert participants in policy formulation or responding passively to structure. For this reason, I also use the interpretive methodology which explicitly incorporates subjective human elements.

This book presents the argument that in addition to being a technical process, urban management is also a political process through which decisions about resource allocation are made. After examining the weight of the evidence in support of this postulate, a moral and ethical framework is suggested for guiding urban policy. Because of the inherent self-interest in humanity, the thesis proposes a structural conflict model in which existing tensions between competing interests are maintained in balance, so as to foster conditions for mutually beneficial decisions in urban management.

Acknowledgements

I would like to thank the following Zambian institutions who provided a wealth of source references - Kitwe City Council, Survey Department, and The National Archives Library. In the UK I made use of the Public Records Office, Zambian High Commission Library, and benefited extensively from the inter-library loan facility at the University of East London. Had it not been for the initial suggestion from Bob Dixon-Gough at the University of East London, this book would probably not have been written.

Particular thanks to the following colleagues: Richard Latham and Doshik Yang each for their unfailing patience as I struggled to produce the maps and other figures and the appropriately named John Spelman of Avalon Associates for his editorial help. Other individuals had also contributed in various ways, in no particular order: Humphrey Chisulo, Nalumino Akakandelwa, Robert Home, Stephen Marshall and Michael Edwards. Finally, I would also like to acknowledge the support and understanding I have received from my wife and children as often I had to work long hours.

List of Abbreviations

ANC	African National Congress
BSA	British South Africa Company
CASLE	Commonwealth Association of Surveying and Land Economy
CBD	Central Business District
CPI	Consumer Price Index
CSO	Central Statistical Office
DBZ	Development Bank of Zambia
DRC	Democratic Republic of Congo
DWSS	Department of Water and Sewerage Services
FiNDP	First National Development Plan
FoNDP	Fourth National Development Plan
GB	Great Britain
GRF	General Rate Fund
GRZ	Government of the Republic of Zambia
GVD	Government Valuation Department
IMF	International Monetary Fund
KCC	Kitwe City Council
KDC	Kitwe District Council
KMB	Kitwe Management/Municipal Board
LWSC	Lusaka Water and Sewerage Company
MMD	Movement for Multiparty Democracy
MTBM	Mine Township Board of Management
NGO	Non-Governmental Organisation
NHA	National Housing Authority
NIPA	National Institute of Public Administration
OECD	Organisation for Economic Co-operation and Development
PRO	Public Records Office
PUSH	Project Urban Self Help
SIZ	Surveyors' Institute of Zambia
SNDP	Second National Development Plan
SOB	South Ore Body
TACH	Total Annual Cost per Household
TAZARA	Tanzania-Zambia Railways
TNDP	Third National Development Plan
TUC	Trade Union Congress
UMP	Urban Management Programme
UNCHS	United Nations Centre for Human Settlements
UNDP	United Nations Development Programme

UNIP	United National Independence Party
USAID	United States Agency for International Development
USSR	Union of Soviet Socialist Republics
ZCCM	Zambia Consolidated Copper Mines
ZNA	Zambia National Archives
ZNBC	Zambia National Broadcasting Corporation
ZNBS	Zambia National Building Society

List of Legislation

Chapter 1

Introduction

The Global Urban Challenge

Current world population is estimated at about six billion people. In relative terms, it is the urban areas which have shown remarkable growth. In 1960, there were 114 cities with a population of more than one million. Of these, 62 were in the developed countries and 52 in the developing countries. The World Bank shows that of the top 32 countries with the highest urban population growth rates, 24 are in the low-income group, seven in the lower middle-income and only one in the upper middle-income, with none from the high-income. Many of the developing world's urban problems are the result of a combination of a number of factors. Any attempt at prescribing solutions must deal with both the factors and consequences. While population growth is one of the factors contributing to the deprivation and decay characteristic of most urban areas in the developing world, there are other factors.

Apart from the demographic and economic factors, the political organizational factor of centralization has concentrated decision-making and with it, resources in the urban areas leading to rural-urban migration. The final factor is one of colonialism. The transfer of foreign social structures and technology, while offering alternatives have dislocated or significantly altered indigenous patterns of development in the developing world. For example, Zambia's Copperbelt is a concentration of towns developed very rapidly in the 1930s, making Zambia one of the most highly urbanized sub-Saharan country (Table 1.1).

The pace of economic growth in the developing world countries has generally lagged behind population growth. In contrast, the industrial revolution provided the needed economic growth for the urbanization of developed countries. Although the rate of urban growth has been on the downward trend since the 1950s, the already existing urban population in most developing countries present a myriad of problems for governments in low-income economies, raising the dated but still relevant question:

> How does a city and its administration obtain financial resources required for public sector activities in a period of rapid growth and economic decline? Can a city keep up with the need for the absorption of large numbers of newcomers, most of whom are poor, even at reduced levels of infrastructure and service provision? (Pasteur, 1978, p.2).

Urban management has, in the last 20 years, become the focus of urban research in the developing world.

Table 1.1   Urban population as percentage of total population, 2000

Country	% of Urban	Country	% of Urban
Angola	34.2	Liberia	44.9
Benin	42.3	Madagascar	29.6
Botswana	50.3	Malawi	24.9
Burkina Faso	18.5	Mali	30.0
Burundi	9.0	Mauritania	57.7
Cameroon	48.9	Mauritius	41.3
Cape Verde	62.2	Mozambique	40.2
Central African	41.2	Namibia	30.9
Chad	23.8	Niger	20.6
Comoros	33.2	Nigeria	44.0
Congo, Democratic	30.3	Rwanda	6.2
Congo, Republic of	62.5	São Tomé and	46.7
Cote d'Ivoire	46.4	Senegal	47.4
Djibouti	83.3	Seychelles	63.8
Equatorial Guinea	48.2	Sierra Leone	36.6
Eritrea	18.7	Somalia	27.5
Ethiopia	17.6	South Africa	50.4
Gabon	81.4	Sudan	36.1
Gambia, The	32.5	Swaziland	26.4
Ghana	38.4	Tanzania	32.9
Guinea	32.8	Togo	33.3
Guinea-Bissau	23.7	Uganda	14.2
Kenya	33.1	**Zambia**	**39.6**
Lesotho	28.0	Zimbabwe	35.3

Source: World Bank, 2002.

The Urban Management Response

As a concept, urban management means different things to different people. To some it is understood in terms of its objectives, to others it is aligned with a system of administration. It follows, therefore, that there is no general definition of urban management. Although there is no single agreed definition, the essence of what is generally accepted as constituting urban management encompasses certain key elements:

- financial, technical and administrative aspects of urban local government;
- deliberate support (public or private) for urban development; and
- promotion of equitable and sustainable urban growth.

Although advanced as a new concept, urban management is an integration of the traditional ideas of planning, with its physical, economic and social concerns, and recently latched to management with its emphasis on efficiency. While planning supposes a control of the sequence of development to some desired end in a predictable social, political and economic environment, the absence of such stability (especially economic) in most developing countries renders the traditional idea of planning alone as a tool for urban management limited. Urban management in most developing countries therefore, rather than being a process of deliberately directing and facilitating urban development, is now mainly about managing the chaotic process of urbanization. Without denying the potential efficacy of public sector planning, urban management is a recognition of the limits of what the state or public sector can do in terms of controlling and guiding urban growth, which control and guidance are implied in planning.

The Nature of Urban Management

As well as being a technical process, urban management is also a political process of negotiation between competing, and sometimes conflicting, demands for limited resources. By political process, it is meant to refer to those processes of human action by which conflict concerning on the one hand the common good and on the other hand the interests of groups is carried out or settled, always involving the use of, or struggle for, power (Banfield, 1964, p.515). Although biased to those processes occurring within government institutions, as understood here, these processes can occur within any human grouping. Batley (1993) supports this political nature by arguing that urban management decides distributive outcomes which include some interests and subordinates others. Devas and Rakodi (1993, p.271) having dispelled the notion of 'public interest', argue that '...city planning and management is an inherently political activity involving choices and conflicts at every stage'.

As an academic discipline, urban management is still in its formative stage encompassing a range of traditional disciplines, for example, town planning; civil engineering, architecture, surveying, economics, law, sociology, public administration, management and others. Because urban management affects people's lives by the way political decisions are made in the distribution of resources and opportunities, such public decisions call for a moral and ethical framework to promote equity.

Potential Weaknesses in the Urban Management Response

Since urban management largely draws from existing traditional ideas of planning and management, it has certain potential weaknesses which derive from its parent ideas.

Firstly, there is the potential danger for urban management to treat all urban problems as technical requiring technical solutions. Implicit in the technical emphasis are the paternalistic views that the professional technocrat in the public sector knows it all and will act in the interest of the public. The technical approach aims to solve the city's problems as if they were purely technical and separated from the wider social and political issues. This technical emphasis manifests itself in an attitude which regards planning and management as mere exercises in optimization. The reality is that the technocrat in the public sector and the state who are supposed to be neutral arbiters between competing demands are themselves motivated by self-interest and will almost always act in ways that benefit their lot unless compelled to act otherwise.

The second weakness arises from a recognition of the first. Realizing that planning is not only a technical process but a political process as well, a paradigm shift in urban research has occurred. Increasingly, many researchers are suffusing politics to all urban problems and therefore seeking political solutions to all problems. The public sphere and policy become the focus of all urban research. The folly of ascribing politics to every urban problem is that it leads people to expect the state to come up with answers to all their problems irrespective of the state's ability. In the process, attention is shifted from the underlying selfish nature of humanity through whose action institutions and policy are created and administered.

The third weakness relates to the dilemma of reconciling public interests and economic efficiency. With the public institutions' failure to deal with service delivery, urban management has latched on to the market approach of privatizing services. The potential weakness and the dilemma this presents, especially in developing countries, is to accommodate the service needs of the poor without compromising economic efficiency criteria in the politically sensitive area of public services.

Central Argument

Having discussed potential weaknesses in current approaches to urban management, the central argument is that, because urban management is both a technical and political process through which decisions on the allocation of resources to competing and sometimes conflicting needs are made, moral and ethical principles are needed to guide decision-making, principles not based on a naive utopianism and egalitarianism nor on unbridled individualism, but on the recognition of the existence in the city of strong poles of self-interest and seek to balance these in a way that promotes a creative tension necessary for the mutual benefit of opposing interests. Given that self-interest (whether the self is a corporate body or an individual) generally determines how these interest groups behave, what can be done to persuade them to act in ways that benefit not only themselves but others as well? This question leads us to a proposal for the maintenance of existing tensions necessary to create conditions for decisions and actions of mutual benefit. The conceptual discussion forming the basis for this proposal is presented in the following chapter.

Aims and Objectives

The overall aim of this book is to analyse the genesis and development of urban management in Zambia and the extent to which local authorities have been able to control and manage development through the provision of services, land and housing. To this end, the following objectives are set:

1. to establish the existence of power bases and what effect these have on the physical growth of a city and its institutions and the ability of all these to perpetuate;
2. to examine the standard and pattern of water services, the extent to which this relates to power bases in the city and the level of economic linkages between communities;
3. to establish, map and account for land and property value patterns and also to determine the relative performance of public and market landed property allocations;
4. to analyse the various modes of housing and ownership patterns as they relate to groups and individuals, determine the level of housing need, establish possible constraints to housing and analyse possible approaches to improve access to land and housing; and
5. to determine the relative distribution of property rates and what impact the inclusion of low-cost housing and the sale of public housing are likely to have on the level and collection of rates.

Relationship to Previous Research

Existing literature on Zambia's urban areas has tended to be confined to colonial urban planning (Conyngham, 1951; Collins, 1969; Gardiner, 1971; Davies, 1972; Collins, 1980; Williams, 1986) or to the examination of colonial institutions (Clay, 1949; Epstein, 1951; Heath, 1953), and analyses of colonial labour policies and social aspects of industrialization (Mitchell, n.d.; Davis, 1933; Mitchell, 1956). Three years after independence, a comprehensive text on Zambia's social geography was published (Kay, 1967). Other themes researched include urbanization and migration (Heisler, 1974; Mwanza, 1979) and settlement planning and policies (Town and Country Planning, 1979; Baloyi, 1983; Rakodi, 1987). Simons' contribution to a reader (Turok, 1979) on development in Zambia addresses a whole range of issues. A detailed survey of informal business in Kitwe and Lusaka was done in 1982 (Haan, 1982). Other studies have focussed on housing among them (Seymour, 1975; Tipple, 1976a, 1976b, 1978, 1981; Collins, 1978; McClain, 1978; Schlyter and Schlyter, 1979; Bamberger *et al.*, 1982; Knauder, 1982; Kasongo and Tipple, 1990). Yet others examine policy, organization and financial aspects of local government (Malik *et al.*, 1974; Simmance, 1974; Pasteur, 1978; Greenwood and Howell, 1980; Chitoshi, 1984; Rakodi, 1988a, 1988b; Mukwena, 1992; Nyirenda, 1994).

While most of the above literature is biased to the treatment of the city as if it were simply a physical structure, other studies extend their perspective to examine the underlying political, economic and social structure which influence the organization of the city's physical structure and the distribution of resources (Seymour, 1975; Rakodi, 1988a, 1988b). Except for one (Deassis and Yikona, 1994), all literature available to this research pre-dates the re-emergence of multi-party politics (1990) and the declaration of Zambia as a Christian nation (1991). Mukwena (1992) discusses local government in 1980. The emphasis in this research is to build on existing theory by providing a comprehensive and up-to-date analysis of the issues leading to principle guidelines based on an informed view of the causal relationships in the issues and also on the right understanding of the nature of the city and humanity.

Structure of the Book

Divided into two, part one of the book comprises three chapters. Chapter 1 is an introductory chapter which sets the theoretical context and global perspective for the whole book. The theoretical area of knowledge related to urban management is explored to reveal the nature and potential weaknesses of the urban management approach to urbanization in the developing world. Having set out the basic argument about what urban management is and a proposal for a conflict approach to the same, research aims and objectives are set out. Chapter 2 presents the philosophical framework for the proposed conflict approach to urban management, the methodology and reasoning process adopted. This is followed in Chapter 3 by a thematic review of literature which captures developmental urbanization issues as well as some thoughts on the study of colonial urbanization.

The second part of the book is mainly a case study of Nkana-Kitwe,[1] the leading mining town and one of the three cities in Zambia. It is intended that

---

[1]   Unless the context suggests otherwise, generally speaking 'Kitwe' refers to the public township and 'Nkana' to the miPning township. When the whole city is intended the double-barrelled name of 'Nkana-Kitwe' is used. Legally speaking, references to Kitwe District or City Council after 1980 should also be taken to include Nkana.

empirical evidence on Nkana-Kitwe will be used to support theoretical statements made to explain antecedent factors in Zambia's urban problems. But before then, Chapter 4 gives the historical and policy context of urbanization and urban management in Zambia. The introductory section to the chapter discusses the historical process of urbanization and associated government attempts to address the issue of growing urban population. Chronologically and thematically organized, the chapter examines the development of urban local government focussing on policy instruments, the relationship of local to central government and the question of financing local government. This is followed by an examination of the development of urban land and housing policies under different regimes. The theory of stabilization and its effect on African housing in the colonial period and the principle of the Durban System in the provision of African urban services are also discussed.

Chapter 5 presents an historical yet analytical discussion of the development of Nkana-Kitwe. Considerable attention is paid to the negotiation between the mining company, the Territorial government and other businesses in the establishment of the public township. The creative nature of the tension that existed between the interest groups is observed. Equally detailed are the discussions on the spatial growth of the town and the evolution of its management structures.

Chapter 6 provides a quantitative, spatial and dynamic aspect on land and property values from 1954, when Kitwe became a municipal council, to about 1995. Aspects of the spatial distribution of land and property values and taxation in relation to local authority revenue are mapped, analysed and discussed. The relation between the public and private supply of land and property is also examined.

As water is one of the basic services which until recently was supplied by both the local authority in Kitwe and the mining company, Chapter 7 is a case study of water supply. This chapter examines the level and standards of supply, the consumption pattern and existing price structure. The city-wide approach in the organization of this whole book enables the identification of, in the case of water, the existence of any economic linkages between different communities, for example, cross-subsidies. The chapter also examines the possibility of using different pricing structures and concludes with a context specific pricing proposal.

Because of the relative importance of the issue of housing and the high demand it makes on urban land not only in Nkana-Kitwe but in Zambia generally, Chapter 8 has singularly been set apart for the study of the development of housing in Nkana-Kitwe. Amply illustrated, the chapter examines the various modes of housing provision. The significance of informal housing initiatives, especially after independence, is also discussed. An attempt is made to estimate current housing need and identify constraints to housing in Nkana-Kitwe using direct and proxy indicators. A comparative analysis of planning standards and existing development is undertaken to examine the possibility of increasing the number of serviced plots by densifying existing settlements and the possibility of reducing plot sizes for new settlements. Recent developments in the privatization of housing are also examined.

The concluding Chapter 9 opens with a re-statement of the introductory argument on what urban management is and proceeds to re-state individual chapter conclusions, linking them to each other and to the main argument. General principles for guiding urban management are suggested. Finally, the theoretical threads of the conflict approach to urban management as postulated in Chapter 2, empirically observed in Nkana-Kitwe, are drawn together into a coherent explanatory theory of urban management in Zambia.

Chapter 2

A Structural Conflict Model
of Urban Management

Introduction

The *International Encyclopaedia of the Social Sciences* defines social conflict as:

> a struggle over values or claims to status, power, and scarce resources, in which the aims of the conflicting parties are not only to gain the desired values but also to neutralise, injure or eliminate their rivals (Coser, 1968, p.232).

The philosophical basis of the proposed conflict approach is Ellul's (1967) penetrating thought 'The Political Illusion'. Ellul observes that the general idea in modern thinking is that tension, arising from conflicting attitudes, motives and values, both at the individual and group level, is unproductive and must be eliminated by adjusting the individual, group and ideologies to the realities of the existing situation. Ellul does not approve of tension for its own sake but only to the extent that it can be surmounted and assimilated. To be of value tension should:

- be a dynamic force capable of resolution on one plane and manifest in a different form on another higher plane;
- not lead to domination;
- be a force between elements of the same system which is capable of withstanding the conflict.

Just as individuals develop in character through their interaction with obstacles, constraints, rules and authority (sources of personal conflict), society also lives and evolves as it learns to deal with tensions in the distribution of economic and political power, but not through adjustment as this results in a false sense of harmony. It is this concept of tension which is taken and applied in the area of urban management.

Conflict and the Urban Domain

One mark of this technological era is that of specialization, this has led to a rather fragmented view of life. We need to make the vital connections which will enable us to have a complete and realistic picture of the nature of the problems we are dealing with. Existing urban management approaches underplay or completely ignore the religious aspect which informs about the nature of humanity, whose welfare they seek.

Socialist thinkers would have us believe that a complete redistribution of resources is possible under prescribed political conditions. The truth is that despite what socialists would have us believe, humans have the inbred propensity to greed and domination. The racial inequalities of the colonial city are the social inequalities of the post-colonial city, this is so despite the socialist rhetoric immediately following the political liberation of colonial cities. Sheppard (1983, p.156) laments, 'It is one of the bitter facts of history that those who have suffered greatly, and know what it is to be victims, so often behave only with self-interest'. On the other hand, the level of inequality in capitalist economies falsify Adam Smith's claim that individuals can be trusted to practise self restraint in their economic pursuits and is proof of the selfish nature of humanity. There is, therefore, a need to have a holistic approach to urban problems, an approach which challenges current thinking on the basis of a proper understanding of the true nature of the city and humanity. Although paradigm shifts taking account of the wider non-material issues have occurred in Neo-Marxist and Neo-Classical thought (Almond, 1993, 1995) the exact nature of the interaction between politics and religion is still a problem for political science (Haynes, 1995).

The success of this approach depends on the relative balance of power between major interest groups. Here is a concept based on the understanding of the true and selfish nature of humanity and the city as a domain of conflict. The realism in this concept is that it makes no pretensions about the true nature of humanity, nor the level of communality/integration, but it also recognizes that society is a shared endeavour. The concept does not introduce tension in the city, it simply recognizes the existence of potentially destructive and competing poles of interest and advocates their maintenance so as to provide a creative tension which has the potential to ensure actions and decisions of mutual benefit. Notice too that this approach does not envisage progress to the romantic situation of complete harmony, but rather it conceives a situation in which resolution on certain conflict points results in conflict on others situated at another plane (Figure 2.1).

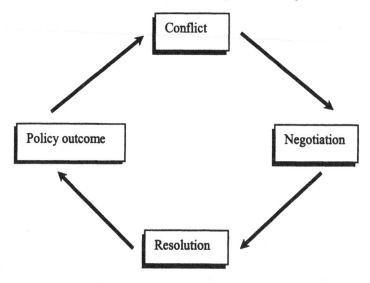

Figure 2.1 Idealized cyclical nature of the conflict approach

It is obvious, therefore, that fundamental to this approach is the presence of powerful interest groups, themselves made up of strong individual interests, otherwise there would be no conflict or tension between the groups. In the context of a local economy and state, if the people are going to be a strong group able to articulate their needs and able to make an effective contribution to the development of the city, then they ought to be treated with dignity both as individuals and as communities. Tied housing, state or private monopoly in the distribution of resources curtail individual freedom and dignity. If individuals are expected to be members of a strong interest group able to apply pressure, then they must be free from the intimidatory forces of the state and/or capital. While acknowledging the individual's freedom and dignity, this approach seeks to balance these with those of the community by emphasizing the responsibility of individuals to others. At this level, we are dealing with the conflict between individual interests and those of the group. There is nothing illegitimate in pursuing individual interests, except that these have to be balanced with those of others. It is important that there is this balance; if not, the individual's enjoyment of exclusive rights to, say, land and housing, would be limited and threatened by those without. Society is the shared enterprise of individuals. Hampden-Turner (1991) and Trompenaars (1993) configure the situation reconciling strong individual interests with those of the group, without the assimilation of one into the other.

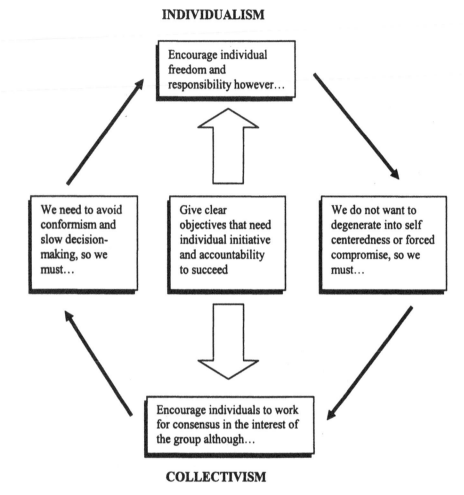

INDIVIDUALISM

Encourage individual
freedom and
responsibility however...

We need to avoid
conformism and
slow decision-
making, so we
must...

Give clear
objectives that need
individual initiative
and accountability
to succeed

We do not want to
degenerate into self
centeredness or forced
compromise, so we
must...

Encourage individuals to work
for consensus in the interest of
the group although...

COLLECTIVISM

Source: Trompenaars, 1993.

Figure 2.2   Reconciling individualism and collectivism

Methodology

Humans, unlike the physical world, have subjective experiences, values, attitudes,
emotions and ideas. All these are real and have meaning to the subjects. While
urban management partly deals with the spatial arrangement of physical
development, decisions on such arrangements are made by human subjects and
affect other humans. For this reason, the 'versten' (German for empathy) tradition
late in the 19th century developed the interpretavistic or humanistic methodology,
arguing that while positivism with its objective claims is suited to the study of the

natural sciences, this methodology is not suited to the study of subjective human aspects (Hughes, 1990).

Urban management, although sprawling across many academic disciplines, is nevertheless focussed on the examination of the public institutional side of urban government (state-centred approach) and community participation (non-governmental approach). While the community advocacy research approach is becoming increasingly popular, it is still valid to pursue a state-centred research framework, because not only does the sense of community decrease (in terms of shared values and interests) with increased territory (a limitation when dealing with city-wide management), but as Okpala (1997, p.1746) observes, 'the most properly functioning, productive and healthiest parts of developing countries' cities today are still those professionally and rationally planned and developed by the public-authority-backed planning system'. Friedmann (1992), arguing for the importance of the state in development, says that although the state may need to be more responsive and accountable, it is nevertheless an important participant in the empowering of poor people. For the above reasons, there is a strong bias in this research to examining state policy responses in the area of urban management. The focus of urban management on institutional and community aspects necessitates the inclusion into the methodological framework of Healey and Barrett's (1990) institutional approach, which will help to understand and explain the relationship between state institutions (local and central) and how private individuals and institutions relate to state apparatus and to each other and generally the complicity of politics in the whole domain of urban management.

Majchrzak (1984, p.18) advances five characteristics which distinguish policy research in general from other research enterprises. Policy research:

- is multi-dimensional in focus;
- uses an empirico-inductive research orientation;
- incorporates the future as well as the past;
- responds to study users; and
- explicitly incorporates values.

Urban policy research, particularly on land, housing, infrastructure and general local governance, sprawls across so many different disciplines. In addition, each of these disciplinary concerns has social, political, technical and economic aspects. A preoccupation with the empirical might address the technical and, to a limited extent, economic issues of this research, but it is an unbalanced treatment of the subject which might lead to the position in which only technical solutions are proposed to all societal problems. Such a focus will invariably lead to a weakness, not only in understanding the causal relationships in current problems and in policy recommendations (Schluter and Ashcroft, 1989), but also in the resulting explanatory theory on urban management.

Referring specifically to urban research, Rabinovitz (1973) argues that urban research is never politically neutral; subjective elements of the researcher's political position are likely to impede on the research. This subjective intrusion should not only be restricted to political ideas, but can be extended to all other ideas and to values, norms, culture and morals. With regard to the colonial legacy of urbanization in most developing countries, value judgements about the kind of people the indigenous were and what kind of society colonial powers wanted to create profoundly influenced policy decisions on planning and housing. Policy making is all about value judgements. If policy is seen to be punitive, it is because policy makers have embedded into it their ideas of normative behaviour, with which they seek to encourage using policy instruments.

Whichever methodology is used, the reasoning process is usually divided into deductive and inductive forms of argument. The former attempts to draw conclusions from a set of assumptions or hypothesis by a logical development of the stated assumptions. Inductive reasoning, on the other hand, involves the verification of theoretical statements by empirical observation (Heywood, 1994). The inductive form of reasoning, which has wide application in policy research, is also sometimes referred to as empirico-inductive as opposed to the hypothetico-deductive. While it is possible in interpretative thought to establish probable relationships between the colonial past and its surviving institutions by studying recorded history and artefacts, colonial value systems and resulting policy and law, such knowledge cannot easily be deciphered using the mechanistic hypothetico-deductive reasoning process because it does not allow for the inclusion of subjective value-judgements.

Although the interpretavist approach helps to establish probable connections between colonial/post-colonial values and the form of institutions established for the management of urban areas, it does not allow for the examination of the underlying political, economic structures and repositories of power which are major factors in resource allocation. Because any proposal to influence the nature of decisions in urban management has the potential to disturb underlying political, economic and social structure, a structural approach, combined with an interpretavistic methodology, enables research to establish probable causal relations, useful in predicting the possible effect and feasibility of policy recommendations on identified interest groups who make up these structures. Recognizing the structural constraints in state policy responses Gilbert and Ward (1985, p.3) advise:

> To understand specific state responses, therefore, a more holistic, class based, political-economy approach is required...most analyses of state policy have ceased to examine the neutral decisions of a technical and objective state; they have begun to examine policies of a state which responds to class conflict and the constraints posed by the international situation... The state is once again regarded as a political entity not as a futuristic, benign and fair-minded arbiter of change.

Because of the disparate nature of urban management research, it has been decided to use an eclectic approach, i.e. one which combines interpretavism and structuralism, but also uses empirical data to support probabilistic explanations.

Because of this research's bias to examine state policy in urban management, the empirico-inductive reasoning process commonly used in policy research enterprise has been used.

Conclusion

Chapters 1 and 2 set the context for this book. The urbanization phenomenon and some of the underlying factors and consequences have been discussed. Urban management has been identified as one of the recent responses to addressing a number of problems in the developing world's urban areas.

The theoretical debate on the genesis, nature and potential weaknesses of urban management has been explored. Although its prominence in urban research is recent, it has been established that urban management is the result of a cross between traditional discourses, and as yet lacks a coherent theoretical core.

It has also been shown that urban management is both a technical and political process through which decisions about resource allocation are made. A number of potential weaknesses in the urban management approach have been discussed. However, it is the allocative aspect of urban management, and the possibility for using allocative criteria narrowly confined to economic and utilitarian principles which have necessitated the proposal for the inclusion of moral and ethical principles to guide decisions on resource allocation. Arising from a recognition of existing self interests which might, at the expense of others, influence resource allocation, a conflict approach proposal has been presented.

As to the choice of methodology, interpretavism and structuralism inform the philosophical argument for choosing the empirico-inductive approach which will be underpinned by empirical data to support theoretical explanations.

Chapter 3

Urban Management Issues: Theoretical Review

Introduction

Without detracting attention from contemporary policy failures in urban management, the historical roots of existing urban forms and institutions for the management of developing country cities have to be properly understood. As to the origins of urbanization in most developing countries, there is no doubt that it has its legacy in colonialism. The first section of this review explores existing literature on the study of colonial urbanism, identifying the main approaches used in studying post-colonial cities and what main findings are drawn out from such studies.

Following the review on colonial cities, issues identified as central to urban management are discussed with a bias to developing countries. Housing is a major claimant of urban land; for this reason, the twin issues of land and housing are taken as one and current literature on their management in the context of urban areas is explored in the second section of the review.

With growing population and diminishing public finances, the provision of public services is another key area in urban management. The third section of this review examines aspects of public/private provision of public services, technological standards, density levels and financial arrangements for cost recovery. Of particular importance to this research are the relevance of indigenous standard arguments against colonial western standards.

One of the recurring reasons given as contributing to the problems of the developing world's urban areas is the weakness of local government. The final section reviews literature related to decentralization - a response to the perceived failure of centralized planning and control.

The Study of Colonial Cities

There seems to be a paucity of published work on colonial cities prior to the Second World War. This could largely be attributed to the fact that most of these cities were still under colonial rule. It is only after their independence (from 1956 onwards) that the colonial cities presented themselves as a testing ground of western theories. The post World War era has therefore seen much research on the colonial city as a laboratory within which various disciplines could test their theories (Fawcett, 1933; King, 1976, 1985, 1990a; Van den Berg, 1984). This section is an attempt to identify the major theoretical frameworks used in the study of colonial cities.

The first framework concerns the study of colonial influences within the city and the lasting impression these have created on the urban form and institutional structures (De Blij, 1968; Kimani, 1972a, 1972b; King, 1976, 1990b; McClain, 1978; Collins, 1980; Haywood, 1985; Ross and Telkamp, 1985). Collins' historical study examines the growth of Lusaka in Zambia from 1931 to 1964. He identifies a shift in influence and control in local government from the civil servants who held sway between 1931 to 1946 to a more independent European settler community between 1946 to 1964, seen for example, in the change in street names from prominent government figures to English Welsh counties. Although Collins finds no evidence of legal instruments used to enforce segregation, he identifies a residential pattern based on racial segregation. It should be mentioned however that the Urban African Housing Ordinance provision (later discussed in Chapter 4) requiring local authorities to provide land or housing for the exclusive use of Africans was positive segregation or possessory segregation. By keeping African wages low, the land acquisition process expensive and minimum building clauses high, Africans were effectively kept out from certain areas where these regulations were selectively applied resulting in exclusive white areas. Commenting briefly on the issue of institutions of government in colonial cities, King (1990b) makes the point that the level of indigenous participation and the predominance of the military, the church or mercantile interests would be dependent on the extent of state and private involvement in the colonization process. Kimani's study of Nairobi examines the spatial structure of land values, seeking to test the applicability of western theories to a developing country city. Using the economic rent theory as developed by Hurd (1924), Haig (1926) and Marshall (1930), he proceeds to test his hypotheses using official data in the city valuation roll. The main conclusion is that western models are relevant to the study of African cities.

The second framework analyses the changing nature and the present role of the colonial city in relation to the global economic system (O'Connor, 1983; Drakakis-Smith, 1987; King, 1990a). Drakakis-Smith's observation is that the period from 1970 onwards has seen a shift to incorporate developing world labour into the international economy by moving manufacturing to the developing world city. The results of this incorporation, Drakakis-Smith (1987) observes, are the continued erosion of local authority autonomy by central government and huge investments in cities so as to attract foreign investment and the simultaneous attempts to rid the towns of informals. In his contribution to the globalization theory, King (1990a) argues that the local urban system is and has been part of a larger system of cities and a result of and a means to understanding the world economic, societal and cultural system.

Christopher (1988) and Home (1997) present a third framework which examines the colonial city in its empire context. Making observations on the extent of colonial city landscape, Christopher (1988) suggests that those colonies where colonial rule reigned for a reasonably long time (at least 100 years for Barbados, Bermuda and Australia) exhibit major impacts compared to Iraq's 15 years of colonialism. He similarly identifies settler colonies as showing more of foreign influence, rather than colonies of exploitation and protectorates which were subject only to British laws but retained a degree of autonomy. Another observation made is that most colonial cities closely resemble the garden cities of England in their spaciousness. The success of the exported garden city concept was affected by varying local conditions as exemplified by Lusaka, Canberra and New Delhi. The most recent, comprehensive and critical assessment of the colonial city at empire level is Home's (1997) *Of Planting and Planning*. Home adopts a thematic approach in which a theme is taken up and followed through in a chronological sense. One of the issues addressed is colonial planning policies which have had a profound impact on the form and function of these cities.

The fourth framework examines the colonial city in its national or regional context. O'Connor's (1983) *The African City* looks at the city in its regional context, concluding that it is important to appreciate the relationship between urban and rural and how one influences the other. He also concedes that the divergence in the cultures of individual cities has not significantly affected the unity in form and structure that is to be found in cities, albeit at differing degrees. O'Connor qualifies the convergence theory of African cities but rebuts the seemingly popular view that all world cities are becoming more and more like the other. Other major points made by O'Connor relate to the overall densities in tropical African cities, which he says are lower than those of western cities. He also observes segregation, sprawl, a decline of land values from the centre and a wedge shaped low-density housing structure.

The study of colonial cities has thus been classified into four frameworks. Of particular relevance to this research are the frameworks which look at the forces within the city and how these affect the spatial structure and distribution of resources, and secondly the globalization framework which examines the changing nature and present role of the post-colonial city in a global economy.

In discussing national urban policy in Chapter 4, this research will allude to the aspects of globalization in relation to housing upgrading policies and recent land policy reform attempts. In tracing the historical development of Nkana-Kitwe in Chapter 5, dominant interests at every stage are identified and particular reference is made to the effect of colonial planning policy on the spatial structure of Nkana-Kitwe.

Land and Housing

The debate on land has largely been dominated by two views. The neo-classical view in its various shades emphasizes the ability of the market to mediate between demand and supply to transfer land to the highest and best use at an equilibrium price (Mattingly, 1993). As a result of its unquestioning belief in the market, state intervention in the market is frowned upon. The market, rather than the state, is considered the most efficient approach to land allocation. This emphasis on the utility and economic concepts of land by the neo-classicals has led to land being regarded as a commodity, a factor of production whose ownership is a source of power and control (Thirkell, 1989). The Marxist view and its derivatives, on the other hand, concerns itself with social economic issues as they relate to land. Questions on who and how land values are created and distributed dominate the Marxist discussions on land (Mattingly, 1993). Because land is a naturally occurring resource, a gift of nature, Marxist tradition argues that in its natural form devoid of human labour, it has no value. If value is the result of human labour expended on the development of land, it is argued under Marxist tradition that this value should be shared with those who create it (Thirkell, 1989).

Consequently, these two views have affected tenurial structures, land valuation, taxation principles and other land management practices. Under the neo-classical regime, the most common form of land holding is likely to be freehold, and taxation is normally based on the capital or rental value as determined by the market. The Marxist view is that land is a national asset and it should be nationalized. Leasehold rather than freehold tenure predominates Marxist-inclined governments, and taxation is likely to be based on normative values rather than real market values (Home *et al.*, 1996). There is also, under the Marxist tradition, a marked emphasis towards public interest in the management of land. Baross and Van der Linden (1990) boldly observe that whatever theoretical perspective, there is agreement that the mere occupation of land transforms it into a marketable commodity. This observation means that, in the final analysis, both mainstream theories have a constrained view of land as a commodity for use, exchange or responsible for structuring socio-economic relations.

Similarly, the discourse on housing, one of the topical issues of urban development in the developing world, has polarized on two traditions. Although they form a united front against capitalist development, they are not agreed on the prescriptions for housing. The Romantic or populist view was born out of the squalor of 19th century industrializing Europe. It is possible to identify some arguments in current debates on housing in the developing world as developments of the populist view. The romanticism lies in the agitation for alternative clean idyllic housing for the worker, a view considered by some as irrational and out of touch with reality (Schlyter, 1984). The populist response to the problem of squalor is good company housing or homeownership.

Arising from the observation of the same squalor during the industrial revolution, Engels (1844), cited by Schlyter (1984) attributes the poor conditions of the working class to competition and repression by the capitalist system which treats labour and housing as commodities subject to market forces. In reference to company housing, Engels argues that this undermines labour power and he also rejects the populist prescription of homeownership as being a capitalist solution or legitimizing the capitalist view of property. In a later discourse on housing, Engels

(1872) argues for the removal of the property owning class. Engels' analysis of working-class conditions in 19th century England is the basis of the Marxist view on housing. Company or tied housing is seen as a means of weakening the bargaining position of the worker who, for the fear of losing a job and house, is likely to moderate his demands. Tied housing in Marxist thought is therefore interpreted as a form of control of labour by capital. Private property or home ownership is seen as encouraging the capitalist view that housing (and labour) are marketable goods.

Having set the broad ideological background to land and housing, the following sections examine institutional, technical and legislative issues which impinge on their supply and the economics and politics of urban land and property.

Institutional, Technical and Legislative Issues in Land and Housing

The prevalence of informal settlements in most developing countries is often attributed to the failure of allocative mechanisms to deliver land for housing to the entire cross-section of communities. Institutional, technical and legislative arrangements are identified as contributing to the failure in the land delivery process.

Three problems are often mentioned in relation to institutional failure. First, existing institutions in most developing countries are said to be either colonial relics or have been borrowed from the developed countries. Critics argue that these institutions were originally set up in a context of slow and controlled urban growth or a highly developed system of administration (Fourie, n.d.; UNCHS (Habitat), 1990; De Soto, 1993; Mattingly, 1993). Given the high rate of urban growth in the developing countries coupled with the limited administrative technology, the formal land allocation system has failed to meet the demand, hence the rise of informality in land acquisition. The second criticism is attributed to the multiplicity of institutions dealing with formal allocation of land (Dunkerley, 1983; Payne, 1989; Farvacque and McAuslan, 1992; De Soto, 1993). It is often argued that no one institution owns the full powers to allocate land, but that allocation is an aggregate of decisions made by different institutions. The third criticism, a seeming contradiction to the second, cites highly centralized institutions as a hindrance to access to land (Farvacque and McAuslan, 1992; De Soto, 1993). It is reasonable to presume in the light of the second criticism, that the centralization being referred to is geographical. The decentralization argument (Durand-Lasserve, 1996; Mutale, 1996; UNCHS (Habitat), 1996b) is for locally-based land administration systems, and there is therefore value in highly centralized but very local institutions which are able to provide a one stop shop for land allocation.

Some of the technical issues related to the improvement of land administration are overly prescriptive, expensive and exacting method-oriented cadastral surveys (UNCHS (Habitat)), 1990). Complicated land registration procedures and land information systems can only be understood by a limited number of people (Fourie, n.d.). The failure of legislation to promote good land administration is also attributed to the use of a multiplicity of inherited or adopted legislative instruments, which often result in conflicts between pieces of legislation unable to deal with the current realities of developing countries (Fourie, n.d.; Rakodi, n.d.; Durkerley, 1983; Farvacque and McAuslan, 1992; De Soto, 1993). Other issues arising from the use of foreign legal systems relate to the conflict between customary and western tenure (Baross and Van der Linden, 1990; Mutale, 1993). Statutes reflecting western value systems may not be understood and/or respected by a people with a different understanding of the concept of property. Apart from the failure (arising from cultural differences) of legislation to promote land management, Baross and Van der Linden (1990) refer to political upheaval following decolonization or revolution as tending to disrupt or weaken state power designed to protect private rights.

Economics of Urban Property

Although traditional African society has historically desisted from treating land as a saleable commodity, the influence of western concepts of land, and increased opportunity value of land due to rapid urbanization have led to a growing acceptance of the use of economic criteria in land allocation (Baross and Van der Linden, 1990; Durand-Lasserve, 1990). There is therefore, increasingly a convergence of thought between mainstream theoretical perspectives and the customary concept that land or its use are marketable goods.

There are two main institutions for the distribution of land and property - the market and the government, the dominance of one or the other being dependent upon government policy. Western capitalist states are more likely to let the market determine the distribution of these resources. On the other hand, socialist inclined states are likely to let government institutions control the allocation process. In reality, however, the convergence alluded to above means that the supply of land and property in most countries across the political spectrum is an outcome of market processes, conditioned and mediated by public sector intervention (Rakodi, n.d.). It is generally agreed that property markets are not as efficient (Rakodi, n.d.; Balchin and Kieve, 1985; Farvacque and McAuslan, 1992) and hence the need for state intervention to mitigate the undesirable effects of market failure, for example, lack of land for housing and high land prices. The level of intervention will differ from one nation to another. Advocates of state intervention present the need to collect windfall gain, ensure land for public services, and access to land and housing by the poor. In the more market-oriented economies, nation-states intervene to provide a basis for the exchange of private property rights, protect private use rights and generally to enforce contractual obligations. For the above reasons, it is not possible to speak of a completely free property market. Increasingly, nation-states in the developing world are encouraging the growth of property markets because these are seen as necessary for development (UN, 1973; De Soto, 1993). The rationale for promoting property markets is that as population increases, commercialization and formalization of property will ensure increases in supply, lower prices and a wider tax base (Rakodi, n.d.; Grimes, 1977; McAuslan, 1985). Others argue, however, that the incorporation of the informal property into the formal market, while enhancing the value of the property, increases rates and rents and might lead to default and dispossession. Payne (1984) suggested that popular attempts at regularizing the informal property market might consolidate those powers (economic and political) which initially were the cause of informality.

Politics of Urban Property

It is generally accepted that the allocation of resources in society is a function of political and economic power (Castells, 1977; Angel *et al.*, 1987; Baross and Van der Linden, 1990). Thus, property distribution in a society does not result from demand and supply based entirely upon the rational decisions of firms and households, but is involved with the political process. The illegal means by which the urban poor acquire land are linked to the state, as nation-states in flagging democracies often use the patrimonial flows of land, housing and public services to ensure social stability, reward political patronage, as sanctions for the lack of political support, and generally to ensure control (Gilbert and Ward, 1985; Sandbrook, 1988; Thirkell, 1989). Thirkell (1989) notes, in addition to rewarding its supporters, a nation-state might use the allocation of land and housing to legitimize its existence among the population, or as a tool to weaken the urban political force by co-opting some urban dwellers into the property-owning class. Baross and Van der Linden (1990) observe that it is in the interest of the government to tolerate informal settlements, because they relieve it from the responsibility to provide shelter and services, so that the provision of basic services to these settlements can be seen as a favour, and not a right. Informal settlers then have the government to thank for the services, eroding their political bargaining strength, an observation also made by Thirkell (1989). Haywood (1986, p.320) argues that 'Political patronage is however a two-edged sword, as a well developed social organization in a popular settlement, with astute community leaders, should be able to exploit it to the advantage of the community'.

Although Menezes (1983) cites the lack of political will to address the land question as being responsible for informal settlements, the problem, according to Baross and Van der Linden (1990), might actually be the opportunity for political control which informality gives. Unless there are other relatively better gains, or there is a genuine desire to grant formal title, no politician would want to lose the opportunity for control informality engenders by regularizing tenure. De Soto (1993), alluding to the high level of political bureaucratic control in the formal land allocation process, claims that a deliberate political decision is needed to allow the formalization of most illegal property titles in the developing world. Mattingly (1993), concurring with De Soto, notes that state intervention in allocating resources has to be a deliberate political action because it may reconfigure the power structure in a society. Since laws are passed by those who already enjoy a measure of political and economic power, why should they compromise their favoured position?

The discussion on the political aspects of urban property suggests a complicity between the state and capital which transcends national boundaries. Thirkell (1989) observes that the state intervenes in land and housing to maintain the reservoir of labour necessary to support the capitalist system. Marxist thinkers claim complicity between nation-states and international capital in the World Bank's upgrading programmes. Schlyter (1984) identifies a number of theories which attempt to explain the real motives of the Bank's efforts in the developing countries. It is clear from the review of the literature above that solutions to the problems of urban property, especially land and housing, in the developing world are not purely technical ones, as Dowall (1991) seems to imply, but involves wider social, political and economic issues. An interpretation of both empirical data and the more elusive state responses offers an appropriate methodological approach in addressing the issues of urban development.

Recent studies show the political factor in the distribution of land and housing. This research will seek to show the existence of this political complicity both at local and national level. In Chapters 4, 6 and 8, we will see the introduction of tied-housing, its effect then and now on home-ownership, and implications on local authority taxation now that all institutional housing is being sold to private individuals. In 1975, the Zambian government intervened to nationalize all land and exert control on all land transactions. We should see in later chapters the reasons for, and the effects of, nationalization on the urban property market. In the meantime, this selected review of the issues in urban management moves to discuss the aspect of the provision of public services.

Infrastructure Services

Netzer (1988) perceives that post-Second World War opinion was that public goods were not being supplied in enough quantities and hence governments in the US and other developed countries were urged to extend their involvement in the public sector. In the 1980s, the pendulum of opinion seems to have swung in the opposite direction with calls for the state to roll back its frontiers, with deregulation and cutbacks in the US. Elsewhere, the state withdrawal can be seen in Britain's privatization and France's decentralization (Alterman, 1988).

Analysing changing attitudes to the role of the state in public services, Gilbert (1992) predicted that by 1995, the drive to roll back the state will have reached its climax, and the call will be for state intervention, regulation and control. All the above cited moves, deregulation and cutbacks, privatization, and decentralization in public services are attempts to minimize government expenditure in the face of increasing financial constraints. If the supply of public services against a background of increasing economic hardship is a global phenomenon, it is worse in the developing countries, which have the added problem of a rapidly growing population against diminishing state finances. These twin problems are identified in most studies on public services provision in developing countries.

Characteristics and Classification of Services

It is generally accepted that two features distinguish public services (and goods) from private. First, the consumption of public services does not diminish their availability to others (the concept of non-rivalry), e.g. air, scenic view and sunshine. Second, there are services which, when provided to one group, cannot be denied to others (the concept of non-excludability), whether or not the consumer pays for the services (Netzer, 1988; Pollit, 1988; Parkin and King, 1992), e.g. the external benefits arising from a good primary health care programme. These characteristics can be contrasted with those of private services, which generate rivalry or competition for their consumption and can be denied or withdrawn from non-payers. It is these characteristics which impinge on the mode of allocation, funding, and organizational arrangements for their provision. People are more likely to pay for private services because of the scarcity brought about by competition and the possibility of exclusion, so private services lend themselves to a market supply. On the other hand, the market is less likely to provide public services because of the difficulties of cost recovery (identification and quantification of individual benefits), and the application of sanctions (excluding defaulters).

Who Provides Services?

Those who argue for the privatization of public services on economic grounds claim that, given the failure of publicly supplied services, the provision of such services can profit from free-market competition or private monopoly. 'Marketable goods and services are better provided by private companies than government, whether the government is local or national' (Carney, 1992, p.211). Carney argues that the nature and objectives of government are incompatible with the efficient and economic supply of goods and services. This argument is countered by Batley (1996), Millward (1988) and Netzer (1988), who observe that the claim of greater efficiency and effectiveness as a result of privatization or contracting out public services is unproven, both in the developed and developing world. A certain level of government participation may be necessary if only to play a regulatory or advocatory role (Carney, 1992; Gilbert, 1992; Cotton and Tayler, 1994; Batley, 1996). Batley (1996) identifies three arguments used to justify government participation in public service provision:

(i)   the public goods argument - because of their collective benefits, cost recovery is difficult and hence the private sector is not disposed to their provision;

(ii)   the market failure argument, due to the tendency to monopoly, and huge capital investments; and

(iii) the equity argument - which advocates everyone's right of access to basic services regardless of their ability or willingness to pay the market price. Given the probability of individuals not knowing what is best for them, the paternal view is that the government should provide these basic services. Notwithstanding general support for a measure of government participation in public service provision, adverse economic conditions and inadequate management structures in the developing world may result in government failure to provide services at a pace which matches population growth, resulting in bias towards the upper-income groups at the expense of the poor (Cotton and Tayler, 1994).

Gilbert (1992), on the other hand, perceives the issue of provision as both political and economic. As a result of economic hardship, there is a convergence of planning approaches between the political Right and Left. The Right considers the withdrawal of the state from service provision in favour of the private sector, as an answer to the problem of sustaining public services in the face of economic decline (Stren, 1991). On the Left, a disillusionment with central command economics and the opening up (Glasnost, Perestroika in the former USSR, economic reforms in China) of these economies to the wider community is advocated (Gilbert, 1992). Developing countries' move to privatize is often imposed by external multinational agencies like the IMF and World Bank, creating a threat to social stability. As more and more government functions are privatized, the form of state control in the developing countries will become more draconian, to subdue a people whose allegiance in former times could have been assured by payments in kind.

Patronage is the glue that holds together unintegrated peasant societies and allows a ruler to govern. Should privatization severely reduce patrimonial flows, governance will necessarily rest more heavily on repression (Sandbrook, 1988, pp.174-5).

Based upon the evidence of seven case studies in Asia, Africa and Latin America, Cotton and Tayler (1994) found an increasing shift from the colonial legacy of state welfare to community provision and management. In another study of six countries in Asia, Africa and Latin America, Batley (1996) identified five main arrangements: public ownership and operation, public ownership and private operation, private ownership and operation, community or user provision, and mixed categories. The developer provision arrangements or exactions discussed in Alterman (1988), and other innovative arrangements in Kirwan (1989), exemplify the range and level of public services which can be provided by the developer in a well-developed market economy. The following section attempts to relate the physical aspects of settlement layout, and density to costs in public service provision.

Spatial Arrangements, Density Levels, Population Growth and Finance

Behrens and Watson (1996) have estimated costs (capital, operating and maintenance) for service provision (water, sanitation, roads, stormwater drains and electricity) at different densities of hypothetical layouts. Relating costs to the geometric layout of settlements, they identified the following important factors: residential density, road and block alignment, and reticulation networks. Based on the (unqualified) assumption that benefits from cost-sharing will be greater than the increase in costs as a result of increased density, they argue that '...even though total internal servicing costs increase with density, the ability to share costs between a greater number of home-buyers results in lower unit costs' (p.240).

A cost analysis (Cotton and Franceys, 1993) based on life-cycle rather than capital cost of a wide range of services and technological levels to ascertain variations in the Total Annual Cost per Household (TACH) relative to plot ratio, settlement layout and plot size, found TACH to be a function of the level of technology, rather than the pattern and physical characteristics of the settlement. The life-cycle cost was discounted to obtain the present value, from which TACH was obtained as a cost indicator. This finding challenges the preoccupation with regularity in settlement planning, and reveals that the greatest savings derive from an appropriate sanitation technology. Using World Bank data on incomes, Cotton and Franceys (1993) claim that in the Philippines the poorest 40 per cent of residents could not afford the high level services on a plot of 30 square metres, and could barely manage lower levels. The use of foreign household expenditure estimates is criticized as a factor in the mismatch between infrastructure standards and cost recovery estimates (Bamberger *et al.*, 1982; Keare and Parris, 1982; Angelo, 1983).

In a study designed to investigate the effect of population growth on the provision of public services, Ladd (1992) used regression analysis for a comprehensive view of the real costs of providing final outputs in public services. Previous work (Wheaton and Schussheim, 1955; Downing and Gustely, 1977) on density and costs of public services is criticized as unrelated to density (Kain, 1967), based on limited assumptions (Windsor, 1979), and failing to account for the wider aspects of population growth and costs of providing final rather than intermediate services (Ladd, 1992). Ladd shows that the planner's assumption of an inverse relationship (the higher the density, the lower the costs), may only be valid for certain densities. The evidence is that 'for moderately populated countries an increase in population density apparently creates a harsher environment for, and thereby raises the cost of, providing public services' (Ladd, 1992, p.292).

On whether population growth can pay the costs of providing public services, Ladd (1992) argues that this depends upon the initial population. Thus, for moderately populated areas, an initial surge in population growth reduces service levels as the local authority adjusts to the growth, and ultimately costs will increase because of the harsh environment created by increased density. For growth to pay its way, therefore, its benefits must be significantly larger, to counter lower service levels and preclude the need for increasing costs to established consumers; the twin effects of surge and a harsh environment respectively. For sparsely populated areas, Ladd (1992) found the effect inconclusive. Increasing the density may reduce costs, but too drastic an increment may have the adverse effects of surge. He also raises the moral issue of who, between established consumers and new ones, should bear the financial burden resulting from growth in population. Traditional approaches ask present users to pay for future service provision and maintenance (Snyder and Stegman, 1987; Cotton and Franceys, 1993). The ideal would be for 'each generation of service recipients and taxpayers [to] pay for taxes equal to the full cost of the capital services it uses' (Ladd, 1992, p.293). The problem of synchronizing these benefits and burdens and the selection of new growth for punitive taxes when established users contribute also to the problem, detracts from achieving such an ideal.

On the fiscal arrangements for cost recovery in public services, Netzer (1988) argues that, to the extent that the private good aspect can be identified and quantified in a public service, that aspect of it should be financed by private user charges based on the marginal cost of provision, and not on the average costs to all users or all units of the service. The argument is that it is not economically efficient to exact charges which do not reflect the marginal cost of providing the service, nor is it equitable to demand payment from people who might not live to enjoy the full life of the service. Users should pay for what they consume, or are expected to consume, and costs should not be transferred across generations. For public services with a public good, Netzer suggests that general taxes or land taxes should be used to finance such services, and examines the suitability of exactions over traditional forms of public finance as a measure of ensuring that the beneficiary pays for the services. Kirwan (1989) observes that various measures have been used by developed country governments in order to reduce the burden of providing public services and improve the financial solvency of local governments: density control (to reduce the need and demand for infrastructure), improvement of existing fiscal policies, cost sharing measures with the private sector, and private financing and control of public services.

Technological Standards and Levels of Provision

The design of public services in the developing countries tends to be based upon traditional western standards, or to experiment with new ideas. Western standards have been criticized as foreign and hence unsympathetic with Africa's current stage of socio-economic, cultural and urban historical development (Okpala, 1987). Reiterating the argument for the use of indigenous standards, Kironde (1992, p.1280) citing examples from Kenya and Tanzania, retorts '...even if resources were abundant, there is still need to question foreign-derived standards, values and attitudes'. The recurring reason for discarding foreign technological standards in preference of local ones is that the former are often unrealistically high and cannot be afforded by urban local authorities and the people. With limited public funds and rising demands for services, the alternative is to lower the standards. The concept of state provision thus constrains the search for alternative means to finance public services. Although he probably enjoys a standard of service higher than the basic he advocates, Kironde offers no evidence of the acceptability of his supposed indigenous standards amongst the population for whom they are intended. One almost detects an anti-foreign tone in the article: 'But even if resources were abundant, there is still need to question foreign-derived standards, values and attitudes' (Kironde, 1992, p.1280). Surely the test should not be whether the concept, or in this case technology, is local or foreign, but whether it enhances the quality of life and is affordable. The basic standards argument should not become an excuse for accepting lower standards or limiting peoples' efforts to achieve higher standards.

Although conventional standards are still in use, a shift towards lower standards results from financial constraints and a desire to be indigenous, e.g. communal water standpipes rather than individual house connections, pit-latrines rather than waterborne sewerage systems. In cases where lower standards have been used, the motivating factor among the professionals and aid agencies has been to provide to the maximum number of people at a minimum cost (Environment and Urbanization, 'Introduction', 1994; Cotton and Tayler, 1994). The perceived wider coverage with lower standards may not necessarily reflect the preferences of the people: contrary to expectations, low-income families may be prepared to pay more to secure a private supply of water (Cairncross, 1990, cited in Environment and Urbanization, 'Introduction', 1994), or for improved sanitation facilities (Altaf and Hughes, 1994). The issue of appropriate standards should not depend upon sentimental arguments about their foreignness, but rather 'How much can be done [with due regard to the people's preferences] to improve infrastructure and service provision to low-income households at a cost that is affordable to government, with substantial cost recovery' (Environment and Urbanization, 1994, p.7).

The debate on the provision of public services, both in the developed and developing world, has one common background - diminishing public financial resources. Chapter 6 addresses property rates in Kitwe as a source of revenue for financing public services. Chapter 7 on water supply, examines levels of provision, existing cost recovery arrangements and level of subsidy, and an alternative pricing policy for Kitwe.

Local Government

Local government's social, economic and political roles in developing countries may be limited because it is weak, inefficient and unrepresentative (Environment and Urbanization, 'Introduction', 1991). The specific roles and relationships with central government and the local community are still problematic, and decentralization has been topical in many countries since the 1980s. Its rise to prominence in many developing countries has been largely due to the failure of highly centralized planning and control of the 1950s and 1960s, the organizational needs of growth-with-equity policies of the 1970s, and lastly in response to devolution pressures with the growth of these countries' populations (Cheema and Rondinelli, 1983). Although a host of potential benefits are advanced as reasons for decentralizing government (Rondinelli, 1981), the process has not been very successful because of a number of constraints.

Decentralization is defined as 'the transfer of planning, decision-making, or administrative authority from the central government to its field organizations, local governments, or non-governmental organizations' (Cheema and Rondinelli, 1983, p.18). Four major forms can be identified (Cheema and Rondinelli, 1983).

(i)   *Deconcentration:* Instead of merely shifting work from the centre to another government office in an outlying area, deconcentration involves a redistribution of functions with the authority to make decisions.
(ii) *Delegation to semi-autonomous or para-statal organizations:* This transfers both function and authority to another agent, not necessarily under a government structure, usually in a specific area, e.g. housing, water, electricity etc.
(iii) *Devolution:* This approach is similar to (ii) except the agents to which responsibilities and autonomy are devolved are local units of government. The existing local government in Zambia meets most of the characteristics of devolution as outlined by Cheema and Rondinelli (1983).
(iv) *Transfer from government to non-government institutions:* The transfer of responsibility from government to some private agency, perhaps NGOs, e.g. limited functions of regulation, licensing or supervision.

Fundamental to decentralization are the following (Environment and Urbanization, 1991):

(i)   *Efficiency and effectiveness:* Donor support has recently included assistance in this area, so that funds are managed at least cost to sponsoring organizations. An example of this interest is the ten-year tripartite Urban Management Programme (UMP) began in 1986 by UNDP, UN-Habitat and World Bank. The long-term hope is that institutional capacity building will not only improve providing and maintaining infrastructure, but will also strengthen institutions that can stimulate and support private economic investment (Clarke, 1991).
(ii) *Political obstacles to municipal reform:* Successful local government might challenge the legitimacy of central government. While central governments may decentralize responsibilities without matching resources or power to local authorities and thus redirect potential criticism, a complete transfer of responsibilities with full powers to attract investment and raise revenue is rare (Velasquez, 1991).

(iii) *Participation and local solutions:* Because the modern structure of local government is often regarded as foreign and inappropriate for most post-colonial developing countries, some commentators (Bubba and Lamba, 1991; Lee-Smith and Stren, 1991; Stren, 1991 and others) have urged a new perspective on local government, moving away from a state-centred approach to participatory structures embracing the community and accountable to local citizens. Herzer and Pirez (1991) and Velasquez (1991) explore in detail participatory arrangements in Latin America.

Mabongunje (1992) supports the community participation approach, contrasting the 'dynamic city-creating activities of civil societies' in the developing countries with the failure of state-centred institutions to guide these activities, and proposes an institutional arrangement whereby the local population can interact with the management of the city. Friedmann's (1992) Alternative Development theory posits a similar, community-centred approach, but with the state as a major player. The most commonly used arguments for community participation are that it improves chances of management success; ensures continuity; promotes a sense of community pride, ownership and self-esteem; increases a sense of responsibility and improves cost recovery. As far as public services are concerned, the fundamental problem with community involvement, however, is one of consensus and standardization. Transactional structures are needed to reconcile competing opinions and form consensus. While decentralization and community participation ideas remain in the urban management discourse because of their potential good, central government inclination to amass and hold on to power constrain their full achievement.

In theory, the legal and organizational structure for local government in Zambia offers potential for decentralization, with considerable powers of taxation, planning and decision-making in many areas of social and economic development devolved to local authorities. In reality, however, real power still vests in central government. Chapter 4 discusses the development of Zambia's urban management institutions, from the Village Management Boards to the contemporary city council, highlighting the changing relations between central government and local government, and the level of political interference in local government. Chapter 5 examines the implementation of national policy at local level in Nkana-Kitwe, looking at the form and functions of local authority. Rates, an important source of local authority revenue, are discussed in Chapter 6.

Conclusion

Having reviewed literature on colonial cities and urban management issues in the developing world, we will investigate in subsequent chapters these issues for the case of Nkana-Kitwe and Zambia. Although many problems related to land, housing, public services and local governance in developing world cities are attributable to present economic constraints, policy failures and institutional weaknesses, this book will explore the contribution of paternal colonialism to problems of centralization of power, poor cost recovery from service delivery, poor levels of home-ownership, and the quality of urban management. The next chapter explores the development of Zambian national urban policy in relation to the issues reviewed here.

Chapter 4

Evolution of Urban Management and Settlement Policies in Zambia

Introduction

This chapter traces the genesis and development of Zambia's urban management from company rule which commenced in the 1890s, through to 2000. Policies, legal enactments and institutional structures, related to urban land policy and urban management, are critically reviewed.

The opening section gives a general country profile, an historical overview of the urbanization process in Zambia and a brief discussion on government responses to urbanization. The remainder of the chapter is organized around selected topics already identified as key to urban management, each tracked chronologically.

The first topic taken up is local government; its development from company rule to date. Considerable attention is given to aspects of democratic representation, finance, functions and the nature of the relationship between central and local government. The post-colonial socialist philosophy of Humanism resulted in significant changes to local government, which are explored. The development of a parallel form of local government by the mining companies is also discussed.

The second topic on urban land discusses tensions between the customary concept of land tenure and colonial land tenure, and differences in tenure granted to indigenous African and European communities. As with local government, the socialist ideology resulted in changes to Zambia's land policy in 1975, which are critically analysed. In 1991, Zambia abandoned the socialist path in preference of a free-market economy, and the debate on urban management and policy/legislation to match the new political dispensation is reviewed, with the role of the international donor community.

The closing section of this chapter is on housing, examining, chronologically, stabilization of labour as it relates to housing, site-and-service schemes and public housing, policy on integrated housing, squatter upgrading and the sale of public housing.

An interpretive methodology has been applied to the relationships between values, ideologies and urban management policies in Zambia, combined with an institutional approach allowing the study of local government structure, a medium for the implementation of such policies.

The chronological sequential framework allows comparisons between different regimes, bringing out changes or continuities in policy. This chapter draws upon selected historical texts, conference proceedings, government pronouncements, national development plans and legal instruments, supplemented by some archival material from both Zambian and British record offices.

Country Profile

Geographically located between latitudes 8-18 degrees south and longitudes 22-34 degrees east, Zambia is a landlocked country enclosing an area of approximately 753,000 square kilometres. Its population has tripled in 36 years since independence from 3.4 million in 1964 to 10.1 million in 2000, with an overall density of 13 persons per square kilometre. About 40 per cent of Zambia's population lives in urban areas, whose most recent growth rate was estimated at 2.7 per cent between 1990-2000 (World Bank, 2002). Zambia is divided administratively into nine provinces: Central, Lusaka, Copperbelt, Eastern, Luapula, Northern, North-Western, Southern and Western. The main economic activity is copper mining, providing about 85 per cent of Zambia's export earnings.

After limited activity by Arab traders and European explorers, significant European penetration into Zambia was made possible by David Livingstone's reports on his visits between 1851 and 1873. The British colonial government, rather than assume direct control of these territories in southern and central Africa, was willing to charter companies and let them occupy, administer and exploit these territories in return for royalties, the British South Africa company (BSA) being one such. Through a number of concessions made directly with African chiefs or by its representatives between 1889 and 1909, the company had a monopoly in land and mineral rights throughout Zambia. While most land rights were surrendered to the British colonial government in 1923, the company held on to mineral rights for over 60 years until a few hours before Zambia's independence in 1964 when it was forced to transfer the same to the Zambian government. Ndulo (1987) provides a detailed historical discussion to the granting of these concessions and later challenges to their validity. The country was administered by the BSA company as two separate entities, North-Western Rhodesia and North-Eastern Rhodesia until 1911, when the two were consolidated to form Northern Rhodesia, but still under company rule until 1924.

In 1924, the British colonial government assumed a protectorate role over the territory until the formation of the Federation of Rhodesia and Nyasaland in 1953, comprising Northern Rhodesia (Zambia), Southern Rhodesia (Zimbabwe) and Nyasaland (Malawi). Nationalist pressure, especially in Northern Rhodesia and Nyasaland, brought an end to the federation in 1963, and Northern Rhodesia became independent Zambia in 1964; the United National Independence Party (UNIP) formed the first Zambian government headed by Kenneth Kaunda.

Zambia's first republic (1964-72) was based upon a plural democracy and a socialist inclined philosophy of Humanism (Kaunda, 1967). While still pursuing Humanism, in 1972, the constitution was amended to make the country a one-party 'participatory democracy', often referred to as the second republic (1972-90). The constitution was further amended in 1990 to allow for the re-introduction of multi-party democracy. In 1991, Kaunda's UNIP lost the election to the Movement for Multi-party Democracy (MMD) led by Fredrick Chiluba, which formed a government. The period after 1991 is referred to as the third republic. Unlike the socialist-inclined UNIP, MMD claims to be a democratic party and strongly believes in private enterprise; state business enterprises established or nationalized during the UNIP era have been privatized or re-privatized, most notably mining interests. Possible effects of the re-privatization of mining interests in relation to urban management are discussed later under the various themes of taxation on property and public services.

A History of Urbanization

This research focuses upon the large towns with populations of over 50,000, whose economic activity is largely industrial, commercial or service. According to the 1990 census, this comprises ten towns mostly along the line of rail (GRZ, 1990). The origin of large scale urbanization in Zambia can be linked to the discovery of copper on the Copperbelt and the subsequent extension of the railway line from South Africa to the Copperbelt. Livingstone, Choma, Kafue, Ndola, Lusaka and the Copperbelt towns are some of the important commercial, industrial and administrative centres on the line of rail (Figure 4.1). The most notable urbanization process in Zambia is the development of the Copperbelt towns in the early 1930s. A deliberate imposition of a hut tax and the attraction of exotic material possessions and lifestyle, forced men to migrate to the Copperbelt and other towns to sell their labour. By independence in 1964, Zambia's urban population was over 20 per cent of the total population (Table 4.1).

Population figures in Table 4.1 show high urban growth rates in the period 1963 to 1969, attributed to the abolition of movement control measures at independence in 1964. After 1969, urban growth rates began to decline with a sharp fall after 1980. The urban population proportion peaked in 1980, declined slightly in 1990 before showing a small rise in 2000. Urban growth rate between 1980 and 1990 is lower than the rural and national average indicating a momentary reversal of the pre-1980 trend. This reversal could be attributed to a number of factors, among them economic and social. The last decade (1990-2000) has seen a slight increase in the urban growth rate compared to the 1980-1990 decade. The urban proportion of Zambia's population at 40 per cent is higher than most countries in Africa (Table 1.1).

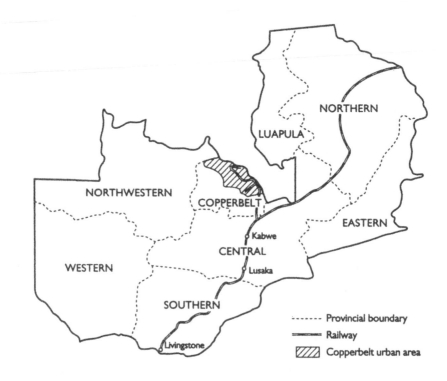

Source: Fergusson, 1999, p.4.

Figure 4.1 Map of Zambia

Table 4.1   Urban and rural population growth trends, 1963-2000

Year	Rural	Urban	Total	Urban as % of total
Population				
1963	2,774,914	715,256	3,490,170	20.5
1969	2,864,579	1,192,116	4,056,995	29.4
1980	3,403,232	2,258,569	5,661,801	39.9
1990	4,477,814	2,905,283	7,383,097	39.4
2000	6,100,400	3,999,600	10,100,000	39.6
Annual Growth Rate				
1963-69	0.6	8.9	2.5	
1969-80	1.6	6.7	3.1	
1980-90	2.8	2.5	2.7	
1990-2000	--	2.7	2.7	

Source: GRZ, 1995a; World Bank, 2002.

Colonial responses to the growing urban population were a combination of movement control measures, such as were used in South Africa: identification certificates (chitupa), tax receipts, visitors' passes, police raids (chipekeni) etc. The declared policy of balanced stabilization, aimed at neutralizing the urban pull factors through rural investment, was never fully implemented. Any success in checking urban population growth was lost when, at independence, movement control measures were removed. Between 1963 and 1969, the urban population grew at a rate 15 times higher than the rural population (Table 4.1). In order to control this influx, an attempt was made in 1977 to return the unemployed to their villages, but opposition from the trade unions blocked this move (Dewar *et al.*, 1982).

Political rhetoric such as 'Back to the Land' has been used alongside some rural development projects, like the rural reconstruction centres established in 1975. Although the above measures seem to have failed in controlling the growth of the urban population, after 1980 the urban population has been in decline. The World Bank's deputy residential representative in Zambia attributed the trend to the difficulties faced by urban residents in accessing services such as health, education and employment (*The Post*, 6 April 1998). As the formal employment sector has declined and subsidies on food and services have been withdrawn, people outside formal employment, rather than take the full impact of economic prices, have chosen to retire to their villages or smaller towns.

The Development of Local Government

Several independent African tribes inhabited the area before the European colonization (Turok, 1989). Although a hierarchical structure of political organization exists in some Zambian tribes (Nyirenda, 1994), with Gluckman *et al.*, (1949) providing one of the earliest studies on the political organization of indigenous tribes in Zambia, modern local government in Zambia was a foreign imposition combining previously separate tribes and clans. In defining the concept of modern local government Simmance (1974, p.4) notes: 'By modern local government, we mean the performance of functions and the provision of services by representative local authorities to which these have been transferred from central government by law'. In another explanation, Nyirenda (1994, p.74) observes that 'local government entails a system where local communities are responsible for their political, socio-economic and development management through a multi-faceted system of democratic process'.

Modern local governments serve the functions of translating central government's policy to locally identifiable territorial units, provision of community and individual services and conveying local community needs to central government. Another aspect of local government in Zambia is a claim to democracy or representative government, important for the social cohesion of culturally mixed urban populations.

Local Government from Company Rule Onwards

For the preservation of peace and good governance (a requirement enshrined in the Royal Charter granted to the BSA company by the British Crown for the administration of Northern Rhodesia in 1889), Village Management Rules were drafted in 1913 to be administered by Village Management Boards appointed by the company administrator. Later these boards comprised elected and nominated members, with most Copperbelt boards, the prime area of interest for the mining company, biased towards nomination (Northern Rhodesia, 1950). This aspect of control by the company continued even after the colonial government assumed control of the territory in 1924. European rate-payers were allowed to elect one member, the other two were the District Commissioner and his nominee. The first such board was set up at Livingstone, Lusaka established a board in 1913 and Broken Hill (now Kabwe) in 1915. These management boards remained the only form of local government until 1927, when the Municipal Corporations Ordinance provided for councils elected by European rate-payers and also including up to three government officials appointed by the Governor. The justification for limiting voting rights to property owners was that they had the greatest stake in the council because of their rates contribution. Municipal councils were empowered to make bye-laws (subject to approval of the Governor-in-Council) and to control African locations within their territorial jurisdiction. Such municipalities were set up at Livingstone (1928), Ndola (1932), with Kabwe, Lusaka, Kitwe, Luanshya, Mufulira and Chingola being upgraded from township status administered by management boards to municipal councils in the early 1950s. In 1929 the Village Management Proclamation was repealed and the Townships Ordinance enacted which provided for the constitution of township management boards to run the more important urban townships (Marquard, 1933). Township management boards were appointed by the Governor and included at least two government officials, while non-official members were, in practice, drawn from the European community. The latter could, from 1952, elect one or more representatives. Town Managers or Location Superintendents were employed for day-to-day administration. Duties of management boards included lighting, sanitation, roads, and production of estimates and accounts, subject to approval by the Governor who also made local regulations regarded as being beyond the ability of the board. Rents and part of the local fines and grants were the main source of revenue for management boards. The theory was that as a township grew in importance it could be elevated to a municipality, and control would pass from the board to a council. From 1929 to 1965, urban local authority in Zambia's public townships was governed by the two ordinances: the Municipal Corporations Ordinance and the Township Ordinance.

Meanwhile, the rapid development of copper mining towns from the early 1930s led to the introduction of another form of local authority outside the municipal and township councils' administration, the private mining township management boards. The obligation to provide worker housing and a desire to stay free of local government control, led to the development of private mining townships between 1932 and 1935. The Mine Township Ordinance of 1932 allowed for the establishment of a Mine Township Board of Management (MTBM), appointed by the Governor from persons nominated by the mining companies. The MTBM had powers to make, amend and revoke bye-laws affecting the mine township, subject to the Governor's approval. Compound managers were also appointed (subject to the approval of the Secretary for Native Affairs) to run the African mine employees' compounds. Although the MTBM had similar powers and responsibilities as those of boards/councils for public townships, it did not collect rates or personal levies and had no allegiance to a political constituency, its officers being nominated by the mining company and not elected by residents (Greenwood and Howell, 1980). The level of company control in the mine townships continued the structure of village management boards on the Copperbelt, which were largely constituted of nominated company officials.

Four interest groups were involved in the development of local government. These were the mining companies, Territorial government, indigenous Africans, and settler traders and farmers, largely of South African origin. The latter not only affected local urban issues, but were instrumental in the formation, in 1953, of the federation of the two Rhodesias (Southern and Northern, now Zimbabwe and Zambia respectively) and Nyasaland (Malawi). A government memorandum stating the paramountcy of native interests over those of immigrant races (Turok, 1989) was challenged by this group, forcing the government to accept a revised wording, that 'the interests of the overwhelming majority of the natives should not be subordinated to those of the immigrant minorities' (H.M. Government, 1932, para.73, cited in Tipple, 1978).

Because voting rights in public townships were limited to the property owning and rate-paying residents (mostly European), the fast growing African population had no direct representation. The Eccles Commission (Northern Rhodesia, 1944) noted this racial imbalance in municipal councils and management boards and urged African representation. A submission to the Legislative Council by the African Representative Council in 1951 for the inclusion of Africans in municipal councils and township management boards did not bear fruit until June 1956, when a committee was appointed to encourage African participation in local government (Northern Rhodesia, 1957a). Even those management boards (Twapia in Ndola; Fisenge in Luanshya; Kasompe in Chingola; Chibuluma in Kitwe, and Kansuswa in Mufulira), which had been established in 1944 to administer African towns under the Townships Ordinance and were expected to educate the Africans in the principles of urban local government, were denied full powers until 1949 (Conyngham, 1951). A commission reviewing the constitution of Rhodesia and Nyasaland advised with respect to urban local government:

> The present system whereby electors for municipal and other local councils are limited to rate-payers results in inadequate representation of African interests. Even where African representatives are elected, this is at present limited to special advisory bodies. We think that Africans should be able to qualify for the vote in

municipal and other elections. We also think that they should be directly represented on the councils themselves and not merely on advisory bodies associated with them (Monckton Report, cited in Malik *et al.*, 1974).

The UNIP manifesto of 1961 demanded a universal franchise in urban local government elections. As a result of sustained pressure, universal suffrage and parity of representation were provisionally provided in the local government elections of 1963 and consolidated as Election Regulations in 1966 (Malik *et al.*, 1974). Although this was a welcome progression from the erstwhile undemocratic local government, it is at this point that Zambian local government elections became politicized. Malik *et al.* (1974) note that except for one council, where the rival African National Congress (ANC) had a majority, UNIP dominated all local councils. Before then, European councillors had been elected as independents.

Immediately after independence, Municipal and Township Ordinances were repealed and replaced by the Local Government Act of 1965. This provided for one-tier representative councils headed by an elected mayor or chairman (depending on the territorial status) assisted by elected councillors. This democratic local government structure, based on universal franchise, did not last long because an amendment to the Act in 1970 empowered the Local Government Minister to appoint mayors. This amendment was the beginning of the erosion of a fully democratic, representative and reasonably autonomous local authority structure. In addition to this amendment, the decision to appoint district governors as party political heads in the district, all in the name of decentralizing the party and government administrative structure, later led to the integration of local authority and the political party UNIP. Under the 1965 Act, local government activities were classified into two groups, those community services financed from the general rate fund (GRF) (e.g. preventive health, refuse collection and street lighting) and those special services meant to be self-financing (e.g. rental housing, water supply, sewage and liquor undertaking). The Act also provided for the appointment of standing and occasional committees in charge of specific activities.

Although the 1965 Act (until 1970) provided for a democratically elected local authority, it has been criticized for failing to ensure direct participation by the public in development issues, except by proxy through elected councillors, and directly only when objecting to council proposals (Rakodi, 1988b). Tordoff (1980) highlights failings:

> ...the generally low calibre of officers who lack the boldness to direct council committees; delays in arriving at decisions as matters are passed between council and committee; limited skills of part-time councillors; and the possibility of political patronage.

In what was justified as a logical progress in the implementation of Zambia's philosophy of Humanism, earlier administrative changes made to decentralize the party and government structures were enacted as the 1980 Local Administration Act, providing for party and government (local and central) structures to work more closely than hitherto (NIPA, 1981). Although laudable in its intent, the strong presence of the party was often a source of tension between the professional district secretariat and the mostly political district council, which will be evidenced from the Kitwe case. The 1980 Act replaced city, municipal, township and mine township councils with the secretariat or executive committee (professional administrative and technical staff) and semi-elected district councils, comprising the politically-appointed District Governor as Chairman; District Political Secretary and two district Trustees; all chairmen of Ward committees in the district elected from party approved members; all members of Parliament in the district; representatives of mass organizations, trade unions, security forces and traditional chiefs in the district (NIPA, 1981). The district council had wide-ranging functions including:

(i)   the formulation and administration of district development programmes in the political, economic, social and cultural, scientific and technological, and defence and security fields;
(ii) the preparation of annual estimates, accounts and reports; and
(iii) ensuring the smooth operation of all public institutions and parastatals in the district, and serving as the final authority in all matters relating to local administration (Mukwena, 1992).

The Local Administration Act of 1980 has been criticized for the lack of a safeguarded legal right to direct participation by the public in development issues (Rakodi, 1988b) and its failure to show how the intended integration between the local and central government was to be achieved, and how this was to affect their respective functions and organizational structures (Lungu, 1986). The dominance of the party in the district councils contributed to their poor performance as funds were diverted to party projects (Mukwena, 1992). Notwithstanding their power to prepare annual estimates of revenue and expenditure, most district councils' progress to greater financial solvency was curtailed by central government and parliamentary power of approval (Chitoshi, 1984). Although the 1980 Act provided for the integration of mine townships into adjacent district councils, the former have continued to be administered by the mines. It remains to be seen how this will work out after the ongoing re-privatization of the mines is completed. This delay in integration has been attributed to problems with staff and property transfers, resistance by mine township residents to paying for hitherto subsidized services (Mukwena, 1992) and the refusal by mining concerns to have their townships incorporated without a corresponding transfer of its social responsibilities to public local authorities (Mbao, 1987). Partial integration has, however, occurred in that mine township residents can now be represented in local government, where before they had no vote in local government.

With the return to plural politics in 1991, local government was harmonized with the new political climate. The Local Government Act of 1991 uncoupled the party structure and operations from the district councils, and moved towards autonomous local authorities with elected councillors and mayors. The councillors sitting in council formulate policy which is implemented by the town clerk's office. Among the functions of the local authorities listed in the 1991 Act are: general administration; agriculture; community development; public amenities; education; public health and order; and sanitation and drainage. The Act replaced the one-tier district council of the 1980 Act with three tiers comprising city councils, municipal councils and district councils. The 1991 Act failed to grant full autonomy to local authorities, as evidenced in central government's power to shuffle senior executive officers, control of finances and presidential directives to local authorities to dispose of their housing stock (a subject returned to later in this chapter). Notwithstanding the change in political parties from UNIP to MMD in 1991, party attitudes remain the same, as will be shown in references to specific issues involving the party and local authorities in urban management.

Local Authority Finance

While some funds for colonial local authorities came from central government, another source of funds was commercial liquor undertaking. As a commercial venture, liquor undertaking was developed in Durban, South Africa, and introduced to Zambia (then Northern Rhodesia) in 1945 as one of the recommendations for financing African housing (Jameson, 1945). The so-called Durban System financed African social welfare services from the profits of beer sales to Africans. Local authorities outlawed the brewing of native beer and set up their own breweries which supplied local authority-owned beer halls in African compounds. Profits from the beer halls went to the social welfare account and were used for housing and other social services. Swanson (1976) discusses the origins and development of the Durban System. Liquor undertaking continued in Zambia until the late 1970s, as this protest illustrates.

> I am a self-employed carpenter and could not get a municipal house as there are none empty and even so the rents are too high for my pocket, having five children. So I built our house with burnt brick and thatch all myself at Kalingalinga where we are not supposed to stay but are allowed because there is nowhere else for us. We have no water, no lights, no schools and no lavatories but the municipals say that they will build us a big beer hall. If we drink beer to give profits then one day there will be water supplied to us in taps. How much beer must I drink before my children can drink water? Do other countries make poor people drink beer to collect money for water? (cited in Hall, 1964, p.135).

Following the Eccles Commission (Northern Rhodesia, 1944), African housing was combined with the Department of Local Government, and the new Department of Local Government and African Housing was transferred from the Chief Secretary's office in November 1947, headed by a commissioner. Although matters of local government and town planning were delegated to the commissioner (in line with the policy of decentralization), financial control was still retained by the Financial

Secretary in the Chief Secretary's office, under s15 and s16 of the Townships Ordinance and s26 of the Municipal Ordinance. Central government control over local government was not confined to the post-colonial period, but had roots in the colonial government. In 1948, the government discussed a decentralization policy to devolve power from the Chief Secretary's office to the Commissioner for Local Government (ZNA: SEC 1/1519: 29 July 1948). The argument used for the retention of financial control was that most local authority funds were received from central government, and therefore it was only proper that such funds be controlled by central government, while allowing for autonomy in the disbursement of local authorities' own revenue (ZNA: SEC 1/1519: 5 August 1948). The department, in 1965, became the Ministry of Local Government, with the power to approve various local authority estimates, capital expenditure and borrowing.

The main sources of local government revenue are property rates and personal levies. Income from property rates increased steadily between 1966 and 1967 and sharply between 1967 and 1968 (Malik *et al.*, 1974). Income from rates was the largest single source of revenue for financing community services between 1972 and 1976 (Rakodi, 1988b). Due to a combination of factors, such as central government control and local administrative problems, income from rates has since declined. The contribution of personal levy to the community service fund was estimated at 10-20 per cent in the period from 1972 to 1976 (Rakodi, 1988b), and central government control limited the full exploitation of this source of revenue through the requirement of parliamentary approval (as provided for in the Personal Levy (Amendment) Act of 1968). Other local government sources of revenue are from licence fees and charges (although the latter is passed on to central government), amounting to 20 per cent of revenue for urban community services in 1974 (Malik *et al.*, 1974). Rentals from council housing have historically been uneconomic due to increased maintenance costs, rent and rate subsidies and profit transfers from councils' liquor sales to the housing account (Malik *et al.*, 1974). With the sale of council houses scheduled for completion at the end of December 1998 (*Times of Zambia*, 1 September 1998), councils have to rely on the new individual owners to pay economic rates on their property.

Because of diminishing central government grants, local authority commercial activities expanded to embrace ventures such as clothing, transport, rest houses, farming etc., but profits were limited by corruption, inefficiency and strict government price controls, and by 1977 most commercial undertaking had been abandoned (Rakodi, 1988b). Until 1972, urban local authorities could lease out state land and collect rents under the head-lease system as provided by the Crown Grant Ordinance (Mushota, 1993), but the withdrawal of head-leases and inception of direct leases denied councils this form of income. To make up for this loss, Land Development Funds have been established in each local authority, to which a percentage of the land rent received by central government for land in the area is paid, but they have limited chances of success because these funds are supposed to come from central government which has been known to act unilaterally in the disbursement of grants to local authorities. A newly introduced source of income for the recovery of expenditure on roads is the fuel levy introduced in the early 1990s. Grants from central government, although contributing eight and 13 per cent to urban councils' community service expenditure in 1965 and 1967 respectively, proved an unreliable source of income because of central government's own liquidity problems: in 1974 central government, without consulting local authorities, reduced the grant in lieu of rates by 33 per cent (Malik *et al.*, 1974) and three years later warned that only 80 per cent of the government grant might be paid (*Zambia Daily Mail*, 29 August 1977).

The main sources of income for large urban authorities are derived locally, but are constrained by the continued financial control exercised from central government, which has denied local autonomy to improve finances by expanding the tax base or introducing economic charges and rates. The country's desire to establish national unity and rationalize the use of limited resources after independence perpetuated a centrally controlled system of resource allocation, yet there appears no justification for the continued existence of central financial control on the basis that 'these are central government funds' especially for large local authorities whose revenue is mainly derived from local sources.

Although the 'rationalization' and 'national unity' arguments for centralization seemed well founded, it became increasingly clear that if local authorities were to operate effectively, then they needed greater financial and legislative freedom. While the 1980 Local Administration Act provided a potentially strong financial base for district councils, progress was limited by the retention of central government and Parliamentary control exercised in many ways e.g. the approval of district councils' annual estimates and expenditure (Rakodi, 1988b). In 1991, the MMD government gained power and immediately re-introduced a measure of democracy in local government, which had been curtailed by the 1980 Local Administration Act. In spite of MMD's rhetoric on decentralization, local authority financial legislative control still vests in the Minister of Local Government and Housing, who controls local authorities through policy guidelines. This lack of financial autonomy was clearly seen in the way in which local authorities, in 1994, were directed by the president to dispose of their housing stock at prices largely influenced, if not dictated, by central government.

In summary, this section has traced the development of local government in Zambia, through five periods. The first period, of village management boards operating under rules drafted by the BSA. This form of administration represented the only form of local government and continued for three years after the end of company rule in 1924.

From 1927 to 1965, three different structures of urban local government existed in Zambia - municipal, township and mine township councils, each governed by its respective legislation.

Between 1965 and 1980, two legislative codes co-existed, one for the public townships and the other for mine townships. Decentralized party and government administration from 1965 culminated in the enactment of the 1980 Local Administration Act, which unified the form of urban local government, previously split between the mine township councils and public township councils (the latter constituting both municipal [including city] council and township council). While party politics penetrated local government from 1963, local government, until about 1970, was fairly democratic and independent of the party structure. Two changes in 1969 and 1970 progressively led to the erosion of autonomy and democracy: the appointment of district governors as political heads and the amendment made in 1970 to the 1965 Act providing for appointed mayors. The period between 1980 and 1991 was a period in which local authority structures became directly linked to the party.

The final period, since 1991, sees the Local Government Act of 1991 re-establishing the universal democratic suffrage of the 1965 Act. Although democracy has been re-introduced in the election of councillors, mayors are elected from among the councillors without the direct participation of the public.

While legislative and structural changes have occurred in Zambia's local government, certain aspects continue from the colonial era. In spite of the decentralization programme and restored democracy, local authorities were, and are still, strongly under central government control. The other continuity is in the area of party politics. Before 1963, councils and management boards were elected or appointed independent of their political affiliation. From 1963, councillors were and are still mostly elected on party tickets.

Underlying value systems and ideologies are at the root of most of these policy formulations between different regimes. This influence on policy is not limited to local government but extends to other areas such as land and housing, examined in the following pages.

Urban Land Policies

Customary Concepts of Land Tenure

Although legal instruments and administrative procedures regulating urban land in Zambia borrow from English land law and reflect western values, they were imported to a situation where the indigenous group's understanding of land tenure was based on custom, after all, 'each society develops its own cultural attitudes to its land' (Green, 1993, p.6). Thus, both customary and western tenure informs the development of Zambia's urban areas. The United Nations' comprehensive definition of customary tenure is adopted, and that is:

> 'The right to use or dispose of use rights over land which rest neither on the exercise of brute force, nor evidence of rights guaranteed by government statute, but on the fact that they are recognised as legitimate by the community, the rules governing the acquisition and transmission of those rights being usually explicit and generally known though not normally recorded in writing' (UN, 1966, p.165).

Customary tenure and its characteristics, according to Bullard (1993), has merits. A committee appointed to examine the systems of land tenure prevailing in the native areas of Northern Rhodesia reported:

> As far as it is possible to generalise, native land tenure in Northern Rhodesia, excluding Barotseland (now Western Province), can be described as communal ownership by the tribe vested in the chief, coupled with an intensely individual system of usage. Every individual member of the tribe has the right to as much arable land as he needs for himself and his family so long as he is making use of this land he enjoys absolute legal security of tenure... The European conception of individual ownership of land has no part in the traditional system of Africa land tenure (Mvunga, 1982, p.16).

In his impressions of indigenous tenure Mackenzie-Kennedy (a Native Commissioner) notes:

> While agricultural lands were allocated to individuals, pastoral tracts were used communally... Unoccupied waste lands were held for the tribe and rights to grazing, timber and forest produce, hunting and fishing rights, the right to use water and manufacture salt, the right to in fact all nature's gifts, were held in common (Mvunga, 1982, p.13).

Attempting to clarify the government position on land, Kaunda states:

> Land obviously, must remain the property of the state today. This in no way departs from heritage. Land was never bought. It came to belong to individuals through usage and passage of time. Even then the chief and the elders had overall control although... this was done on behalf of the people (Kaunda, 1968, cited in Mvunga, 1982, p.14).

Thus customary tenure in Zambia enshrines the concepts of: communal ownership of all pastoral rights and natural products of the land; private use of arable land; prudence; security; and the concept that bare land has no exchange value.

Land Tenure and Policy Under Company Rule

Before the BSA company divested its control of Northern Rhodesia in 1924, almost all land occupied by Europeans in the territory was held under freehold or leasehold. In 1905, North-Eastern Rhodesia produced the first regulations governing the registration of land (under exclusive use-title) and North-Western Rhodesia followed in 1910. In 1911, the two Rhodesias combined to form Northern Rhodesia. On instructions from the British colonial government and in order to safeguard their own property, land was set aside for the use of indigenous Africans. A comprehensive Land Registry was proclaimed on 1st November 1914 to provide for the registration of documents granting or dealing with interest in land (Fairweather, 1931b). Although an amendment in 1944 was meant to convert the system from the registration of deeds to the registration of title, in practice little changed. The enabling legislation is still referred to as the Land and Deeds Registry Act and land owners still have title deeds issued to them.

Colonial and Federal Land Tenure and Policy

In 1924 control of the territory passed from BSA company to the British colonial government. Although the 1911 Northern Rhodesian Order-in-Council instructed the BSA company to set aside land for the use of indigenous African people, this was only formalized by the 1928 Order-in-Council which provided for two categories of land: Crown land (administered by English land law for European settlers); and reserve land for indigenous Africans (administered by customary law).

Because of overcrowding in the reserve lands, the 1947 Native Trust Order in Council set apart a new category of land (Trust Land) made up of all unassigned land, forest and game land, and unutilized Crown land. This third land category was also set apart for the use of indigenous Africans (Roth *et al.*, 1995). What was designated as Crown land is a narrow corridor of land astride the railway line from Livingstone to the Copperbelt, with a few isolated pockets in major provincial settlements. This research focuses on Crown land (now State land). Fairweather (1931a) gives a detailed account of the several forms of tenure available to Europeans. By 1931, tenure for township plots was mostly granted as long leaseholds to Europeans and tenure for Africans was on comparatively short leases, reflecting the semi-permanent housing structures affordable by most Africans and also the colonial attitude of a temporary urban presence for Africans. In 1950, Lusaka allocated 52 such short leases (10-15 years) in designated African areas, 30 of which had been developed by 1951. Security of tenure depended upon meeting obligations of ground rent, service charges and 'good behaviour' (Location Superintendents' and Welfare Officers' Conference, 13-15 February, 1951. Filed at ZNA: SEC 1/1568).

To support this exclusive allocation, use, and commodification of land under company and colonial rule, a fixed cadastre system was established, linked to the South African system (Dare and Mutale, 1997) and operated by a profession of cadastral surveyors trained in the measurement of land using elaborate methods and precise equipment (Home and Jackson, 1997). The establishment of land registries, supported by diagrams framed from ground surveys and beacons on the ground, provided for the definition of exclusive rights and the extinction of what indigenous rights pre-existed in the land. New elements introduced were: private rights which before were limited to the use of arable land; insecurity, because the exclusive nature of rights led to the label of squatter being applied to those occupying land to which, according to the colonial system of land law, they had no rights; commodification of land in the situation in which bare land could not be sold or mortgaged.

Following recommendations by a committee on urban land tenure (Northern Rhodesia, 1957c), the Crown Grant Ordinance of 1960 (later called the State Grant Act) allowed 99-year leases to convert to freeholds, provided that all requirements relating to the minimum building clause and charges had been complied with. Another provision of the Crown Grant Ordinance was the head lease system, which allowed for big local authorities to sublet land for a lesser period than the 99 years on which they held the head lease. Some of the advantages of this system are the privity of contract between the local authority and the sub-tenant in matters of development control, development charges and service provision. In addition, the head lease provided a source of income for local authorities by way of plot premiums (discussed in detail in Chapter 6) payable by plot holders (Mushota, 1993). With effect from 10 November 1972, the head lease system was abolished and replaced with a direct lease issued by the Commissioner of Lands on behalf of the state. One reason advanced for this change was that the premiums charged by local authorities were too high and thus inflated the price of land (Oke, 1974; Town and Country Planning, 1979).

The abolishment of the head lease constitutes one of the changes made to colonial land policy after independence, this and other changes, and the post-colonial conflicts and accommodations in the administration of land policy, with a customary land tenure testify to the foreignness of some aspects of colonial land policy in Zambia. The next section attempts to examine independent Zambia's land policy, and the extent to which conflicts in colonial land policy have been harmonized with the country's ideological and cultural views.

Post-Colonial Land Policy

Independent Zambia's land policy cannot be divorced from political ideology. Zambia adopted the philosophy of Humanism - a man-centred ideology aimed at the eradication of the exploitation of man by man (Kaunda, 1967, 1974). A comprehensive report (GRZ, 1967) on planning, land tenure and acquisition, housing, squatter settlements and local authority controls was submitted in 1967, but was criticized for its lack of imagination and Anglo-American views (Mvunga, 1982). While accepting the criticism, McClain (1978, p.64), himself a Commissioner, remonstrates '...the most serious handicap the commission faced was not the legal education of the members or their devotion to English law but rather the uncertainty about fundamental policies which the new Zambian government intended to follow'.

Although there might have been doubts about what policies the UNIP government would pursue after independence, its 1959 manifesto is clearly based on the traditional concept of usufruct: 'anyone can have access to and use a piece of land but cannot claim any form of ownership' (Anon, 1993a). *Humanism in Zambia* (Kaunda, 1967) illustrates the underlying political ethos and emphasizes two problems related to land, the holding of undeveloped land by absentee landlords and the dangers of commodification of land. Whereas the earlier UNIP

National Councils (1968, 1969) dwelt more on commercial and industrial activities, the 1970 National Council dealt more directly with land, its concentration and informal settlements (Kaunda, 1970). Conceding that freehold was only two per cent of all forms of tenure, it proposed that:

(i)   all land should vest in the President;
(ii) all freeholds should convert to leaseholds for a minimum period of 100 years;
(iii) land under customary tenure should not convert to leasehold; and
(iv) other reforms directed at improvement of the use of agricultural land.

The 1970 Lands Acquisition Act added powers to the earlier Public Lands Acquisition Ordinance over absentee landlords. Compensation was only to be paid in limited cases, and aggrieved owners could only appeal to the National Assembly and not the High Court. Policy pronouncements were finally drafted and enacted as the Land (Conversion of Titles) Act No. 20 of 1975. In his watershed speech to the Sixth UNIP National Council heralding the 1975 land reforms, then President Kaunda referred to a case in Lusaka in which George Louis Lipschild sold to Solar Investments, on 3 April 1975, three plots (each less than a quarter of an acre) for K150,000 and on the same day, Solar Investments sold one of the plots to Development Bank of Zambia (DBZ) for K100,000. Kaunda referred to this transaction as madness, and directed that Solar Investments return the money to DBZ, and that the undeveloped plot was compulsorily acquired by the state (Kaunda, 1975). The 1975 Act introduced far-reaching changes in urban land tenure and administration by: providing for the vesting of all land in the President by the conversion of all freeholds into leaseholds; abolishing all value transactions in bare land; and introducing the concept of state consent in land and property transactions.

Mvunga (1982) questions the validity of the traditional heritage claim (Kaunda, 1968) used in supporting the 1975 Land Act, the general equation of states with chiefs and headmen with absolute ownership rights to land. Agreeing in part with the concept of undeveloped land having no value, Mvunga states that this fails to recognize external market pressures, and also criticizes the administrative mechanism for enforcing the 1975 Land Act.

Acknowledging problems inherent in the 1975 Land Act, the government, in 1980, appointed a committee which recommended that the 1975 Land Act be repealed as it was restricting the free operation of the market, fuelling the high property values and retarding property investment (Mulenga, 1983). An examination of Zambia's land delivery system (Oke, 1974; Halcrow Fox and Associates, 1989; Mutale, 1993; Bruce *et al.*, 1995) reveals the level of bureaucratic complexity (Figure 4.2) of the land delivery system before changes in 1995, the complexity of which has been compounded by political complicity (*Zambia Daily Mail*, 9 March 1994; *Times of Zambia*, 15 August 1994; *Sunday Times of Zambia*, 14 September 1997; *The Post*, 22 December 1998) and a high level of centralization. Although an attempt at decentralizing the land delivery system has been made by the establishment of a regional lands office in Ndola, this regional office was still in 1998 not yet operational (Hansungule *et al.*, 1998).

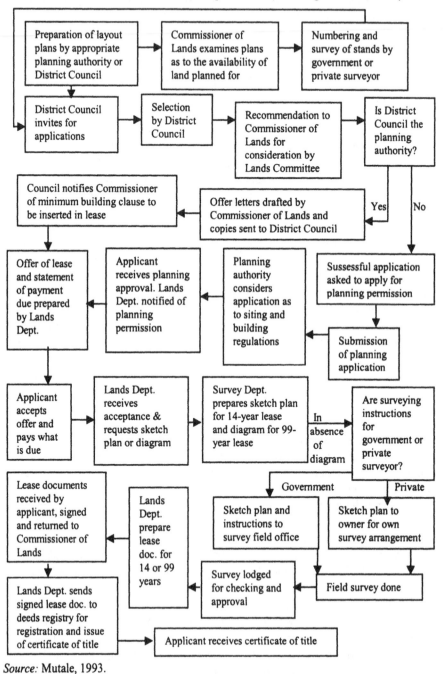

Source: Mutale, 1993.

Figure 4.2 Activity flow of the urban land delivery process in Zambia

Apart from the freeholds granted in farms and plots under company rule, there was, under the colonial and federal system, a measure of continuity of the concept of public ownership akin to the traditional communal ownership and private use rights. Under the colonial/federal system, land was granted as leaseholds in urban areas and permits of occupation for farms, although the latter could eventually graduate to a freehold title after fulfilling certain conditions. This continuity was completely harmonized in 1975, when all freeholds were abolished and converted into leaseholds deriving from the state (equated to chiefs and elders in the customary context). The abolition of all land transactions for value in 1975 accorded with traditional heritage, removing a tension that hitherto existed because of the commodification of land which was widely sold by public auction. There is evidence, however, that traditional attitudes were changing to accepting the exchange value of land (Van den Berg, 1984).

Although the 1975 Land Act aimed to harmonize the country's land policy with tradition and accorded with the philosophy of Humanism, Simons' (1979, p.21) conclusion after analysing Marx and Lenin on private property is that the traditional Zambian private land use system was capitalist in character.

> The Marxist-Leninist position, in summary, is that the relationship between the owners of production resources and the actual producers determines the nature of social structure and the form of the State. If nationalised land is leased to individual producers who sell their commodities on an open market the result is capitalism. If production is carried on collectively by co-operatives or state enterprises in a planned economy the result is socialism. By these criteria the preponderance of individual family farmers and individual homeowners in Zambia's 'mixed' economy gives it a pronounced capitalist orientation.

If Simons' analysis of Marx and Lenin is accurate, then we must accept that the intent of the 1975 Land Act was in conflict with the general capitalist orientation of Zambian society, all be it tempered with the strong socialist character of traditional Zambian life.

In October 1991, the MMD took power. At a seminar organized by the Surveyors' Institute of Zambia (SIZ) on the theme 'Land Development Policies - the need for change', the new Minister of Lands in his opening address reiterated the government's intention to liberalize the land and property market by ending the system of state consent and withdrawing state monopoly in property financing, attaching economic value to land, and decentralizing the land allocation procedures by giving more power to local authorities (Chileshe, 1991). The concept of privatization was paramount to the new government's thinking. In an address to the donor community, the new government declared its intention of dispensing with the 'command economy', seeing itself playing the role of facilitator (GRZ, 1991a).

A major national consultation took place from 19-23 July 1993 on the theme 'land policy and legal reform'. A number of papers, representing views from government, opposition, church, World Bank, and professionals, were presented. In a long paper, prefaced with a historical review of land policy and a scathing attack on the 'cheap and pretentious rhetoric... of the UNIP government', Mushota (1993, p.13) for the government stated that 'we in the MMD believe in and uphold the principle of private ownership of land and shall promote it as a matter of constitutional right'. UNIP (Anon, 1993a) restated its 'strong principled stand of free traditional heritage'. Mushota (1993) asserted that the government, while promoting private ownership, should have the power to regulate the operation of the land market, thus implicitly accepting the principle of stewardship inherent in a leasehold system. Although the government had not explicitly stated its position vis-à-vis freehold tenure, its strong position on privatization was enough to stir a heated political debate. In her address, Zambia's Labour party leader implored Zambians to stand up and fight for their land that was being grabbed from them and sold off to foreigners (*The Weekly Post*, 2-8 October 1992). Another political observer wrote 'they are selling your dignity for generations to come...'. By August 1994, the land issue was already politically charged, and the government's Land Bill was defeated and withdrawn by government.

A close examination of the Land Bill reveals that in many respects it did not differ from the Land Act of 1975. The following were the Land Bill's new points relating to urban land:

(i)   the extension of the concept of economic value to all undeveloped land;
(ii)   the establishment of the Land Development Fund (an idea borrowed from the colonial Orders in Council); and
(iii) the establishment of a Lands Tribunal.

The Land Bill further sought to restrict the number of transactions requiring state consent, and to effect the surrender of land held by local authorities, so that all leaseholds would derive from the state and not as subleases from local authorities (as had been the case under the head lease system). Land held by local authorities under the Housing (Statutory and Improvement Areas) Act was exempt from this provision. Notice that although the head lease principle had been abolished in 1972 (see earlier), local authorities still held land from the state granted before 1972 which they could sublease. Apart from the above changes, the Land Bill, like the 1975 Land Act, provided for the continuation of leasehold alongside customary tenure.

Government's failure to win approval for the Land Bill was largely of its own making, as there had been little popular debate. Meanwhile, the government was meeting with representatives from aid donor institutions and drawing up a time-frame for policy reform, leading the opposition and general population to suspect that government was acting under pressure from donors. After a period of debate and considerable pressure from the World Bank (Palmer, 1998), the Bill was finally enacted into law in 1995.

The Common Leasehold Schemes Act (providing for strata titles) was also enacted in 1995, and the Land and Deeds Registry Act was amended to allow for titles to be registered against descriptions with no cadastral survey. The land surveying profession, in relation to the whole land reform programme, felt sidelined or dragged along by the scruff of the neck, and lobbied hard but without success. The Land Survey Act and Land and Deeds Registry Act are in conflict: the latter includes a description of the land as a basis of registration, but not the former. The anxiety expressed by land surveyors is not without justification. A decision taken by the government in concert with representatives from the World Bank, British Aid and USAID, but without the knowledge of the Survey Control Board of Zambia, was that the survey profession was going to be opened up to persons with qualifications recognized by international bodies. A year earlier, international lawyers, accountants, engineers, architects and surveyors had met under the auspices of the Organization for Economic Co-operation and Development (OECD) and discussed the creation of a more open market for their services (Dale, 1995).

The above section has identified key elements of land policy in different regimes, and how these have changed or have continued with changing political dispensations. Under company rule, land changed from a freely available gift of nature in communal ownership to a commodity for sale and the subject of exclusive ownership, a change which created conflict between the law and a people accustomed to traditional dispensation through chiefs, leading to the by-passing of formal allocation through squatting. The 1975 Land Act sought to harmonize land policy with the traditional system by decreeing that undeveloped land had no value. The re-introduction of the concept of value for undeveloped land and other recent land reforms seems to have been influenced by the international aid community.

National Housing Policies - Past and Present

Company and Colonial Housing Policies

The earliest proclamation on housing in Zambia (then Northern Rhodesia) was made in 1917 (Northern Rhodesia, 1917, cited by Tipple, 1981). The Mines and Sanitation Regulations required mining companies engaging more than 300 African workers to provide housing, and was the basis for subsequent housing legislation. In 1929, five years after the colonial government assumed control of the territory, the Employment of Natives Ordinance required employers to arrange at their own expense accommodation for their employees, but not for their dependents.

Labour Stabilization and African Housing: The theory of stabilization is central to a discussion of colonial housing policies. Until the 1940s, the general theory relating to the African presence in urban areas was that it was temporary and that Africans would retire to their villages at the end of their stay in town. African housing, therefore, was regarded as a receptacle for a transient unskilled labour force. Despite the mining companies' intention to stabilize its labour as early as 1933 (Heisler, 1971), arguing that this would raise productivity and was the moral thing to do, the Territorial government was not keen and argued that an economic slump such as occurred in 1929-30 would have adverse effects on a stabilized out-of-work African population (Notes on the Stabilization of Native Labour, June 1944. Filed at ZNA: SEC 1/1320). Missionary opinion as to whether stabilization for urban Africans was desirable was also divided during the 1930s. Heisler (1971) observes that the Church of England was the last missionary group to give stabilization its blessing. The closure of Nchanga Mine, as a result of a slump coming soon after, vindicated government fears and put a temporary stop to the question of stabilization. This, combined with the labour disturbances on the Copperbelt in 1935, resulted in a sharp swing of opinion in the mining companies who, in seeking to avoid unionization, now preferred a migratory labour force. 'Detribalization and urbanization should not be encouraged in any manner', said the Chairman of Roan Selection Trust in 1937 (Meebelo, 1986, p.166). Following labour disturbances on the Copperbelt in 1935, the Russell Commission appointed to enquire into the disturbances criticized the barrack type of mine housing and urged the Territorial government to stabilize the labour force (Northern Rhodesia, 1935). The government still remained undecided on the question of stabilization (Heisler, 1974). In 1938, the Pim Commission, appointed by the Colonial Office to find ways of reducing government expenditure following the Governor's report of financial hardship in the territory, recognized that *de facto* stabilization was already taking place, but was cautious not to recommend a deliberate policy of stabilization because of the financial implications of providing services to a stabilized population (Heisler, 1974). Following a second wave of strikes on the Copperbelt in 1940, the Forster Commission was appointed and made submissions to the Territorial government in July 1940, although pressure from the mining companies delayed publication until February 1941. A joint statement prepared by the Territorial government and the mining companies before publication of the Forster Report sought to explain what both the mining companies and the government had done to effect the recommendations. It is probably the refusal by government to commit itself to stabilization which prompted the Colonial Office to direct:

> It is clear that Northern Rhodesia copper industry cannot be regarded as other than permanently established, and it follows that the time is approaching when the government's attitude towards the settlement of a permanently industrialised community at the mines must be defined more precisely (ZNA: SEC 1/1320).

In addition to the Colonial Office's directive, a memorandum on the Forster Report was published in London by the Anti-Slavery and Aborigines Protection Society, advocating stabilization (Memorandum by the Anti-Slavery and Aborigines Society, filed at ZNA: SEC 1/1320).

Following a recommendation in the Forster Report, Sir John Waddington (the new governor) appointed Lynn Saffery, a sociologist from the South African Institute of Race Relations, to study, among other aspects of African living conditions on the Copperbelt, stabilization and the resulting need for social services. Using detailed records kept by Scrivener, the compound manager at Nkana Mine, Saffery, made his submissions in February 1943. Although the

Labour Commissioner, R.S. Hudson had, in October 1943, expressed the need for addressing the issue of stabilization in the light of Saffery's findings, on advice from Governor Waddington in March 1944 to the effect that the report should not be published because mining companies were critical of Saffery, the Colonial Office agreed and the report was treated as departmental for internal circulation only (Heisler, 1974).

Largely as a result of pressure following the publication of the Forster Report, a local commission chaired by L.W.G. Eccles, a land surveyor and Commissioner of Lands, investigated the administration of native locations in urban areas. Notwithstanding the Employment of Natives Ordinance of 1929, employers had failed to provide reasonable housing, and Eccles' Report (Northern Rhodesia, 1944) recommended that employers provide housing for workers and their dependents (estimating that at least 10,000 units at a cost of £1m were needed). Eccles also recommended the establishment of a separate African Housing Department, which took place in 1946 (Local Government and African Housing, filed at ZNA: SEC 1/1519). Eccles' findings and recommendations were supported by Walton Jameson, an expert in 'sub-economic' housing from South Africa. With pressure coming from within and without, the Territorial government was instructed in April 1944 by the Colonial Office to prepare a stabilization policy. In January 1945, the Territorial government eventually made its position clear when it stated that:

> ...endeavours should be made to provide a balanced stabilization in the rural as well as urban areas and that rural amenities should be extended in an effort to induce Africans employed in the towns to return to the country on retirement... the declared policy on this matter should be the recognition of a provision for that degree of urban stabilization which exists from time to time and the progressive development of the rural areas to keep pace as far as possible with progress in the urban areas, special provision being made to encourage the retirement of urban workers to rural life with the object of achieving a balanced stabilization in both urban and rural areas' (SEC/LAB/27, cited in Heisler, 1974).

In 1948, the Urban African Housing Ordinance required employers and local authorities to provide housing for African employees and their wives; the emphasis on family housing was a significant progression from previous housing instruments which restricted housing to workers alone, although the principle of tied-housing was still maintained. Because the majority of mine employees on the Copperbelt were men, many women could only get access to housing as wives of mine employees. A marriage certificate was needed as proof of the union (Hansen, 1997). These two forms of housing (employer and local authority), until the recent decision to privatize all institutional housing, have been dominant in Zambia.

'African towns' were also established outside municipal boundaries for the self-employed on the Copperbelt who could not be housed on the basis of existing legislation. Sited about four miles from established municipal areas (see Figure 5.1), five such towns exist on the Copperbelt: Chibuluma in Kitwe, Kansuswa in

Mufulira, Kasompe in Chingola, Twapia in Ndola and Fisenge in Luanshya. Houses would be built by owners on land leased from the local authority, or the local authority built the houses and sold or leased them to individuals. A local administration system operated with limited powers, staffed by Africans under the control and guidance of the Provincial Commissioner through the District office. While the Copperbelt had the benefit of 'African towns' and Mine Townships, the unemployed or self-employed in other towns could only settle either as squatters, or as tenants on private land (derisively referred to as 'Kaffir Farming'). Because of the rapid growth of such settlements, described as 'notoriously insanitary' (Northern Rhodesia, 1950), the Private Locations Ordinance was enacted in 1939 to regulate their growth (PRO: CO/795/1O4/45260: 10 December 1938).

Recognizing the problems in providing council housing for married Africans, the Location Superintendent Conference meeting in Lusaka in 1950 recommended that Africans be allowed to build their own houses in African areas of the municipality on plots leased from the council. The conference observed that this would help alleviate the current housing problem and contribute to the contentment and promotion of civic pride in the African (Location Superintendents' Conference, February-March 1950. Filed at ZNA: SEC 1/1568). Lusaka established a number of such schemes, such as Mapoloto (Plots) in Chilenje, a low-cost home-ownership scheme established in the 1930s. Despite the formation of the Department of African Housing, public African housing remained a problem, and the African Housing Board Ordinance of 1956 constituted a statutory body with wide ranging powers to advise and assist local authorities on matters pertaining to urban African housing (Northern Rhodesia, 1957b).

Progress from the squalor of labour camps for single workers only was possible for three reasons (Heisler, 1974): the shift in public opinion criticizing the colonial government's abdication of its trusteeship duties; the potential for higher productivity and profit resulting from a happy, healthy and smaller work force; to promote social stability, especially as the territory was aspiring to get white self-rule from the Colonial Office. The new willingness to address African social concerns resulted from a shift in government attitude influenced by the Fabian Society, and in response to increased labour agitation on the Copperbelt. The labour strikes of 1935 and 1940 and the formation of African trade union movements, guided by Mr. W.M. Comrie, a TUC advisor sent by the British government in 1947 (Makasa, 1981), gave the mining companies a fright and led to the improvement of housing.

The Zambian contribution in the Second World War is recognized in existing Zambian army barrack names, such as Kohima, Burma and Arrakan, imported from the South-East Asian theatre of war. Zambian nationalist leaders in their mass rallies pointed to how little Africans had gained from their sacrifices (Makasa, 1981), which pressure contributed to the 'Homes for Heroes' allocation of over £10m for African housing in Northern Rhodesia under the Post-War Development and Welfare Plan. African labour agitation should be balanced with the colonial government's trusteeship position in attempting to curb the excesses of mining capitalism when explaining improvements in African mine housing. In spite of progress towards a positive urban policy, the mining corporations were still reluctant to invest in social infrastructure. Sir Ronald Prain (Chairman of Rhodesian Selection Trust) in 1956 stated:

> The issue really turns on the question of housing... The capital outlay required from employers in order to house a rapidly growing African population is frightening; it may tend to make the Copperbelt mines high-cost producers with all the dangers attendant on that, and it is actually inhibiting the development of secondary industries (Prain, 1956, p.307-8).

New African miners' housing in the early 1950s included 60 per cent well serviced family units (Hailey, 1957), largely as a result of pressure from the government and labour agitation, a standard of provision which led Oppenheimer to boast that the mining industry had:

> ...brought large numbers of Africans to conditions where they live in well built brick houses, and enjoy dietary, sanitary and medical services that are as good as modern science can provide. On the mines amenities now being provided for African employees include electric light in their homes and individual water-borne sanitation - facilities which today are not available to about half the population! Never in the world's history has more rapid progress been made by any section of the people' (Meebelo, 1986, p.230).

The improvement in African mine housing also kept African wages low, because of the monopoly power of the mining companies; that these wages were insufficient is borne out by the huge 80 per cent raise awarded to Africans after a successful strike in 1952 (Simons, 1979).

A total of 100,000 houses were built for African occupation by various institutions in the territory between 1948 and 1964 at a cost of at least £32m - 20 per cent of all investment in the territory over that time (Heisler, 1974). White Location Superintendents were appointed to manage African housing areas, and the kind of paternal control exercised is exemplified by a proposal made at a Location Superintendents' Conference in February 1951 to discontinue the showing of cowboy films to Africans because these were at the time considered harmful to African behaviour (ZNA: SEC 1/1568). Figure 8.10 shows a typical local authority African house developed in the territory after the Second World War.

At independence in 1964, Zambia inherited a housing problem. Twenty-one per cent of Lusaka's population lived in unauthorized settlements. Although the mining companies had managed to control the development of such settlements on the Copperbelt, of the 40,000 houses owned by the mining companies in 1966, only 15,000 could be regarded as family housing (two- or three-roomed structures of burnt brick or concrete, with an average floor area of about 38 square metres, private toilets and water facilities); the rest were 'squalid, humiliating and overcrowded' (Simons, 1979). As for the physical layout of the major towns, a segregated pattern of development kept the main racial groups physically separated with separate provision of housing, schools, shopping facilities, administrative centres, and even burial grounds. As far as the Copperbelt was concerned, the mining companies and local authorities each developed their own segregated township.

Post-Colonial Housing Policies

With the removal of movement control measures at independence in 1964, urban population grew rapidly, adding to the housing problem. Table 4.2 illustrates the growth in housing demand in relation to supply in Zambia's three cities from 1966 to 1970.

Table 4.2   Total housing stock and demand for Zambia's three big local authorities, 1966-70

Year	Lusaka	Kitwe	Ndola
1966 HS	11,033	10,434	13,124
1966 WL	14,105	8,442	6,433
1967 HS	12,648	11,168	13,618
1967 WL	13,732	12,865	8,567
1968 HS	12,763	11,286	14,349
1968 WL	13,818	17,588	11,404
1969 HS	12,686	11,718	14,686
1969 WL	16,392	19,277	13,100
1970 HS	13,532	13,726	15,113
1970 WL	21,454	21,479	15,078

Note: HS refers to Housing Stock and WL to Waiting List.
Source: GRZ, 1971.

The establishment and growth of informal settlements reflected housing need. A decade after independence, the percentage of informal housing was 46 per cent in Lusaka, 42 in Kabwe, 36 in Kafue and 32 on the Copperbelt (Simons, 1979). Since an extensive literature exists on post-colonial housing initiatives in Zambia, a brief critical discussion follows.

(i) Site-and-service and low-cost council housing: The first post-colonial housing policy guideline is Circular 17/65 issued on 24 April 1965, outlining government policy to high density housing. With cost limits and minimum standards of accommodation (GRZ, 1965) it introduced the concept of 'site-and-service' which would attract a 50 per cent government subsidy towards the cost of land and services (roads, sewers and water); occupiers were expected to meet the other half of capital expenditure through a 40-year loan scheme. Although home-ownership

was intended, the form of tenure (monthly tenancy) fell short of what would be expected of such a scheme (Collins, 1978). Because of the rapid rise in construction costs, the African Housing Board recommended a lowering of standards in 1966, and the use of direct labour to spread benefits widely (GRZ, 1966a). Circular 59/66 on 'Aided Self-Help Housing and site-and-service schemes' (GRZ, 1966b, cited in Tipple, 1976b), allowed for the use of low-cost materials to maximize participation, and a Land Record Card as proof of ownership. While tenure arrangements in circulars 17/65 and 59/66 have been criticized as deficient, the latter offered a renewable 10-year lease - a considerable improvement in security compared to the monthly tenancy of circular 17/65 (Collins, 1978).

Despite these attempts at site-and-service and rental housing schemes, housing for the urban population still remained a critical problem. Circulars 29/68 and 30/68 (GRZ, 1968a, 1968b, cited in Tipple, 1976a) introduced the idea of basic site-and-service schemes, with services limited to roads and communal water supplies. Believing that these basic services would appeal to many in squatter areas, councils were encouraged to invest in these schemes, but basic services were not a strong enough incentive for resettlement, so strong-handed methods were introduced with the demolition of a number of settlements in 1970-72 (Mulwanda and Mutale, 1994). Between 1966 and 1970, a total of 63,410 housing units were provided under the site-and-service scheme compared to 15,993 complete units built under the public local authority scheme (C.S.O., 1971).

(ii) Integrated housing: At independence, pressure was brought to bear on the Commissioner of Town and Country Planning to abandon the colonial system of residential segregation in favour of an integrated system more sympathetic with the new Zambian political philosophy of Humanism.

In August 1968, the Minister of Local Government and Housing issued a policy statement on integrated housing. A number of families (say 200) would form a cell with all community services provided within, and a collection of cells would form a neighbourhood. Integrated housing, based on the romantic humanistic view of egalitarianism failed because it ignored the real nature of humanity, the individuality of families and communities, and experienced mundane planning difficulties. For example, although sympathetic with the government ideological position which informed residential integration, Kitwe City Council (KCC) had practical engineering difficulties in developing such housing (private communication, 13 June 1994). Only a few settlements are now integrated, mostly between high- and medium-cost houses.

(iii) Squatter upgrading: While local authorities, notably Lusaka, carried out squatter demolitions in 1970, the ruling party denied any involvement (Mulwanda and Mutale, 1994). At a UNIP National Council, the President spoke patronizingly about such settlements:

> We dismiss it as something which has come about as a result of bad planning and selfishness on the part of the colonialists and that it is therefore a legacy of the past.

This is largely true but it does not provide solace to men and women given the label 'squatters' (Kaunda, 1970, cited in McClain, 1978, p.66).

In 1972, the head of Lusaka City Council's Squatter Control Unit was dismissed after the District Governor (a political appointee under the decentralized system of local government) intervened when squatters in Lusaka's Chipata compound resisted a clearance attempt (Mulwanda and Mutale, 1994). In the same year Zambia was declared a one-party state, and UNIP had a reason to protect squatter settlements (in which it had established party branches) as a thank you gesture for the support in establishing a one party state, as well as recognizing the futility of squatter demolition.

> ...although squatters' areas are unplanned, they nevertheless represent assets both in social and financial terms. The areas require planning and services, and that the wholesale demolition of good and bad houses alike is not a practical solution (GRZ, 1971, cited in Collins, 1978, p,116).

On housing policy, the Second National Development Plan (SNDP, 1972-76) recommended that a selected number of squatter settlements be upgraded while continuing with site-and-service schemes. Home-ownership was encouraged by the provision of finance at commercial rates for house construction or purchase, while subsidized rental house-building by larger local authorities was stopped. While the SNDP was conceived during a period of relative prosperity for Zambia, the effects of the unilateral declaration of independence in southern Rhodesia in 1965, followed by world sanctions, put a strain on Zambia's economy. Housing had to compete with the construction of the Tanzania-Zambia Railway (TAZARA) for budgetary allocations during a period of depressed copper prices. As a result, the implementation of the SNDP was delayed and when launched, government's commitment to upgrading was a meagre K5m (£3.5m) compared to K35m (£24.5m) for sites-and-service schemes (Kasongo and Tipple, 1990). A temporary change of policy during the plan period saw the construction of rental houses in large urban areas (Town and Country Planning, 1979). While the government was committed to upgrading, it did not match its words with funds. Government spent 75 per cent of its housing resources on high- and medium-cost houses (Schlyter and Schlyter, 1979), and overspent on institutional housing for middle- and higher-income groups by about 50 per cent (Kasongo and Tipple, 1990). Mulenga (1981) attributes the poor performance of the SNDP housing proposal to poor organization, financial and legal difficulties, lack of qualified personnel and the reluctance of council officials to accept the value of self-help. Seventeen years on, staff and finance were still a problem, at least for Lusaka City Council which had an establishment of 50 posts in the planning department but had only eight staff, and nine staff out of an establishment of 40 in the building inspectorate (*Times of Zambia*, 28 February 1998).

Several legislative enactments passed during the SNDP are relevant to the housing discussion. The Housing (Statutory and Improvement Areas) Act of 1974, enacted a month before the signing of a K26m loan agreement with the World Bank for squatter upgrading in Lusaka, simplified procedures in the land delivery process and planning control by excluding the application of the Lands and Deeds Registry Act, Land Survey Act, Town and Country Planning Act, Rent Act and the Stamp Duty Act to statutory (site-and-service schemes and council housing areas) and improvement (squatter settlements) housing areas. The tenure arrangements offer 30-year leases and occupancy certificates for statutory and improvement areas respectively - a significant improvement over the earlier ten-year leases of circular 59/66. The duration of occupancy certificates was left to the discretion of individual local authorities (Hansungule *et al.*, 1998, Kitwe Evidence). Although a potentially useful piece of legislation to address the problem of registering a huge number of titles in a short span of time, a consultancy report (Halcrow Fox and Associates, 1989) found that the Housing Act of 1974 had not been fully used. While it gives an improvement in tenure security, Mvunga (1982) criticizes the use of the term 'occupancy licence' as this connotes the idea of 'making that lawful which would otherwise be unlawful'. Whereas the Housing Act of 1974 presents a radical shift in government attitude towards squatter settlements, laws still exist preventing unlawful occupation of land, for example, Housing (Statutory and Improvement Areas) Act s39(1); Land and Deeds Registry Act s35; Lands Act 1995 s9(1).

The closest we get to official thinking on upgrading policy at the time is the resolution at a housing seminar attended by officers from central and local government in 1981. Seminar participants resolved that, while employed persons in squatter settlements should be resettled (presumably in site-and-service areas or institutional housing), unemployed squatters should be urged to go back to the land 'so that the unplanned settlements vacated by them are wiped out' (cited in Schlyter, 1984, p.53). Instead of squatter upgrading, the policy was for local authority rental housing, employer housing, individual and private company investment in the formal rental market. A policy of demolition was re-introduced in the 1980s, but couched as squatter relocation (Mulwanda and Mutale, 1994). This official limited view of squatter upgrading is reflected in the Third National Development Plan (TNDP, 1979-83), allocating a tenth of the national housing budget to site-and-service and upgrading, the balance going to institutional housing (GRZ, 1989). In 1991, at least 38 unauthorized houses were demolished in Lusaka's Kanyama compound (*National Mirror*, 16 December 1991). Apart from a decision taken in 1994 to submit to the donor community, a plan to regularize squatter settlements in major cities (which component the World Bank indicated willingness to fund), there has been no real progress with squatter upgrading on the scale of 1970s upgrade in Lusaka.

The return to plural politics has strengthened the squatters' bargaining power, with a higher level of political and legal articulation. In two incidents, squatters in Nkana's Nkandabwe and Ndola's Chichele violently resisted attempts to evict them (*Times of Zambia*, 6 August 1997). In October 1997, 500 squatters in Nkana's Saint Anthony successfully filed a High Court injunction (a legal term made popular during the transformation to multi-party politics) restraining Nkana Division from evicting them from a settlement they had built illegally (*Zambia Daily Mail*, 20 October 1997). In all these cases, party leadership at the grassroots was involved on behalf of squatters, as with UNIP during its tenure. In Lusaka's Kalikiliki compound, over 100 land allocations were made by the MMD, with prices ranging from K2,000 to K90,000 per plot (*The Post*, 14 October 1994). With the changed political atmosphere, the use of force to control squatting would be politically unwise, and inaction might be politically unwise and aesthetically unpleasing. As long as squatters feel secure or have recourse to political manipulation and legal representation when threatened, the situation remains stable. If government cannot house, or fails to enable its citizens to adequately house themselves, and fails to protect private use rights, then its legitimacy may be questioned.

(iv) Housing for all: The housing section of the 1991 MMD manifesto (MMD, 1991) states that the MMD shall:

(i)   aim at promoting the construction of adequate and suitable housing for all with water and sewerage services and, where practicable, electricity. Existing squatter compounds shall be systematically upgraded and where necessary resisted;

(ii)   encourage private organizations to build houses either for sale, or for their staff;

(iii) create an enabling environment where all Zambians will have access to financial resources to build a house;

(iv)   create an appropriate fund by formulating marketable instruments to attract long term investments from pension schemes and other long term sources of funds to finance the housing programme.

A series of meetings and provincial workshops culminating in a national conference on housing were conducted by the Ministry of Local Government and Housing, and a draft housing policy issued in March 1995 whose central goal is

...to see that all its citizens are adequately housed. This will entail the supply of about 110,000 units annually during the next ten years from 1996 to 2006, for all categories of the community in order to meet the housing demand and accord with the United Nations Global Shelter Strategy for All by the year 2010 (GRZ, 1995b, p.13).

The government draft housing policy emphasizes home-ownership through the provision of land and services, discourages institutional housing, and makes guarded reference to upgrading through the use of such language as 'discretional support of self-financing upgrading schemes'. The housing policy launch in 1996 was followed shortly by a declaration of two council housing areas in Lusaka as statutory housing areas, thus allowing for the sale of houses and grant of individual titles to tenants.

(v) Sale of public institutional housing - empowerment or gerrymandering?: In a move described as 'empowerment' (reducing employee dependence on employer), the MMD government (more precisely the President) directed all public institutions (notably local authorities, Zambia Consolidated Copper Mines (ZCCM), and central government) to sell all their housing stock to sitting Zambian tenants. Section 35 of the Employment Act was amended to remove the obligation for employers to provide at their expense, accommodation to their employees and dependents, but not without Trade Union pressure for compensatory pay rises. The Zambia Federation of Employers had been advocating its removal since 1994, pledging to pay employees wages sufficient to house themselves (*Financial Mail*, September-October 1994). The opposition claimed that the government was trying to buy votes by giving away houses which they had not built. The decision was timed close to MMD's end of first term of office, and all housing stock dating before 1960 was given to the tenants freely, except for legal fees which ranged from K10,000 to K20,000 (£5-£10). The government argued that local authorities had already recouped their capital expenditure through rentals. House sales discriminates against those who are not institutionally housed, denying them a lifetime chance to own a house cheaply. Local authority tenants who, since 1959 have paid rents (which are now being taken account of as contributing to the paying off of council expenditure on the same), but had ceased to be a tenant just before the government decision to sell, lost out, as did those tenants who had bought at economic values before the Presidential directive to reduce prices (*Zambia Daily Mail*, 17 June 1996). Although the aspect of empowerment is laudable, it could lead to anarchy and brutality as government attempts to assert control over an empowered population, while those left out of the scheme will be alienated. To re-dress such problems, the government released K3bn (£1.5m) through the Zambia National Building Society (ZNBS) for lending to civil servants who had not been beneficiaries of the sale of pool or council houses (*Times of Zambia*, 9 July 1997).

This section has discussed the development in housing policy. The policy of labour stabilization and its cost implications has been examined. Improvements in urban housing policy, however limited, are seen as a result of this debate on stabilization, labour and nationalist agitation, and other factors, such as changed attitudes toward African welfare, and the potential for improved production on the mines. The colonial legacy of Zambia's housing problems is clear, while the site-and-service response made popular after independence was also a colonial idea to address the African housing problem (although never widely implemented). The discussion has also explored changing official attitudes to squatter settlements after independence, and the complicity of politics in squatter housing issues. Finally, the promotion of home-ownership through the sale of institutional housing marks a departure from tied-housing.

Conclusion

Urban management is both a technical and political process, through which resources are distributed among competing needs by public decisions. At city level, local authorities are the major public institution in the allocation of land or development rights, housing and infrastructure services in most developing countries.

Using the interpretive and institutional approach, the chapter has attempted to identify dominant interests which have affected the development of policy in Zambia's urban administration, and to establish at national level, relationships between the nature of policies and institutional structures in various periods and how they have changed with changing interests, values and ideologies.

The opening section of this chapter examined the origins and development of Zambian urbanization; policies, legal instruments and institutions developed to manage and provide land and housing for the urban population; and government responses to urban growth. The development of local government has been explored through five periods:

- 1913 - 27 Village Management Rules (1913);
- 1927 - 65 Municipal Corporations Ordinance (1927), Townships Ordinance (1929), Mine Township Ordinance (1932);
- 1965 - 80 Local Government Act (1965);
- 1980 - 91 Local Government Administration Act (1980);
- 1991 onwards Local Government Act (1991).

Aspects of democracy, finance, the form and functions of the institutional structures, and their relationship to central government have been explored.

Recognizing the foreign origins of urban land law, the section on land examined the key elements defining customary land tenure in Zambia and identified three central points of discussion: the attaching of exchange value to land and the introduction of the exclusive and registrable ownership rights in the early 1900s, the 1975 Land Act, and recent debates on land reform leading to the Land Act of 1995.

The housing section traced the evolution of housing policies in Zambia from tied housing, the stabilization of African labour and progressive African housing improvement. Notwithstanding this progress, the discussion on company and colonial housing closes on a negative note, citing the presence of informal settlements in Lusaka and squalid housing for many on the Copperbelt. The housing problem after independence was compounded by the freedom of movement and residence enshrined in the constitution. Government responses designed to address this problem have been explored. This section concludes with the privatization of institutional housing, noting both government and opposition claims and commenting upon inequities.

In relation to the central argument in Chapter 1, Chapter 4 investigated sectional interests in Zambia's urbanization process. The first major economic interest is the BSA company, which administered the country from about 1890 to 1924 and had both the administrative autonomy and economic monopoly. Indigenous Africans were excluded from local government under company rule, and their housing limited to the squalid conditions of the labour camps.

In 1924, the British colonial government assumed control of the territory, but remained weak administratively and economically and reliant upon the company's administrative experience. Not until three years later (in 1927) did the colonial government enact its first urban management ordinance. Before then, urban administration was regulated by the Village Management Rules drafted under company rule in 1913. African interests were still subordinate, especially in local government, in which the African remained unenfranchised until 1963.

Political power in 1964 was transferred to the independent Zambian government and, with the nationalization of the economy and land between 1969 and 1975, the state emerged as the major interest. The dominance of political interests in local government, especially after 1980, testify to this.

Changes in urban policy were brought about by a number of factors. Firstly, the shift in colonial attitudes to a more socially responsive urban approach, in part influenced by strong moral principles, but also a response to nationalist agitation. Secondly, the improvement in mine worker housing, although in part motivated by economic self-interest, was also a response to the paternalist colonial government (exercising a trusteeship duty) and labour demands for improved wages and housing.

Before 1924, the BSA company had dominant and unfettered control. After 1924, the colonial government became the countervailing force to the excesses of company domination and capitalism, providing a creative tension and leading to a general improvement in urban policy.

In the next chapter we will see how these power bases practically affected the urban development of Nkana-Kitwe.

Chapter 5

'Let Us Build Us a City'– Copper Mining and the Growth of Nkana-Kitwe

Introduction

This chapter moves the discussion from the general theory and places it into a selected geographical context. The city of Nkana-Kitwe in Zambia has been identified for use as a detailed case study and appropriate references made to the other cities of Lusaka and Ndola. Cogent arguments are presented to support the validity of the choice of the case study area, citing among the arguments the declaration of Zambia as a Christian country and Kitwe's biblical motto.

In its original biblical context, the rallying call 'Come, let us build us a city' (Genesis 11:4) refers to humankind's self-centred attempt to build a city with its Tower of Babel that reached to the heavens. The attempt was, however, a failure. That this defiant, egotistical and failed attempt to build a city should have been chosen by Kitwe's city fathers for the town's motto either denotes their ignorance of the Bible or a determination to succeed where Babel failed. But, there is a hidden irony in the rallying cry, of which those who chose it were probably unaware. The call was made by the builders of the Tower of Babel, who wanted to climb up to Heaven, and God punished them for their presumption, dividing them against each other by making them speak in different tongues. The mining town of Nkana-Kitwe received its fame by delving into the ground in search of wealth, rather than up into the sky in search of glory, and the irony is that the city of Nkana-Kitwe was, from the beginning, a divided town: between the European and African residents, and between the private mining interest and colonial authority. These divisions have affected the physical form and structure of the city through to the present day. Indeed, the very name Nkana-Kitwe reflects its history as 'Twin-townships': Nkana the mining township and Kitwe the public township.

Using both interpretative and institutional approaches, this chapter identifies political and economic structures, explores their power relationships in influencing the urban development, built form, and urban administration in the city of Nkana-Kitwe. The concept of power is a contested one, but for the sake of brevity, it is here taken to mean the ability of institutions or individuals to galvanize around a common interest and to direct or influence urban processes and institutions.

The chapter opens with a justification of why the city of Nkana-Kitwe has been chosen as a case study and is followed by brief discussions on each of the following topics: history of copper mining and the functional evolution of Nkana-

Kitwe, population growth and economic activity. Later on in this chapter, early negotiations over the twin-township structures form the central focus of the chapter. Protracted three-way negotiations involving the mining company, the Territorial government, the Colonial Office in London and other territorial local business interests in the establishment of Nkana-Kitwe are chronologically explored.

Having identified the main power bases in the city's establishment, attention shifts to tracing the physical development of the city and that of its management structures. Using objective references to the city of Nkana-Kitwe, the concluding section draws together into explanatory theory the interests and power structures in urban planning and management - interests and power structures referred to in relation to the central argument and the need for moral principles in urban management (Chapter 1).

Like other chapters, the chronological organization of this chapter is intended to capture the dynamic city processes so as to provide a historical context of the city's urban problems and also to inform the formulation of relevant urban management principles. Although based on a wealth of archival data, the emphasis in this chapter is not to write a history of the city of Nkana-Kitwe, but rather to relate history to the complex of forces which have affected urban development and management in the city.

Selection of Study Area

This research is designed as a longitudinal case study of the development of one Copperbelt town in Zambia (Figure 5.1). One of the key reasons for choosing a Copperbelt town as a case study is that this area has always been seen as central to the urbanization process in Zambia.

The chosen mining town of Nkana-Kitwe (see Figure 5.1) is one of the three cities in Zambia. Kitwe City Council's Bible-based motto 'Let Us Build Us a City' (see Figure 5.3) makes the city particularly relevant as a case study for three reasons. Firstly, the city's motto ties in with a thesis which proposes moral and ethical principles in urban management, and secondly, the motto expresses a deliberate desire to guide and control the development of a town, guidance and control which have become difficult to realize in the face of rapid urban growth and a poor economy. Thirdly, the identification of the 'us' in the city's rallying call to development, enables this research to focus on the colonial origins of urbanization in Zambia and its legacy on the existing physical form and institutional structures for urban management in Zambia. The longitudinal nature of the research design provides a historical and dynamic context in which the process of urbanization and related issues can be studied.

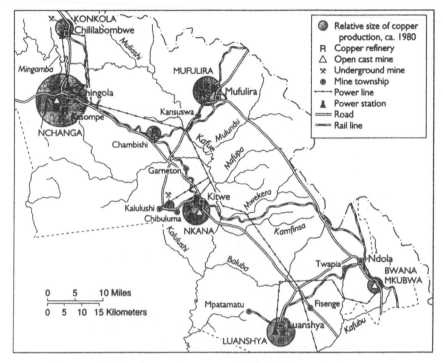

Source: Ferguson, 1995, p.5.

Figure 5.1 Map of Zambian Copperbelt

While the motto of the city as bequeathed to it by the colonial city administration has not changed, the original heraldic device (Figure 5.2) on the city's coat of arms was changed in 1969 to one which accurately conveyed the statement portraying Nkana-Kitwe as an independent copper mining town. The crown was replaced by two picks - symbolizing the city's mining activity. The heraldic beasts either side of the coat of arms were replaced by two African men in bush-jackets holding T-squares. The presence of two technocrats on the coat of arms signifies a determination by Kitwe City Council to continue with the rational development of the city. The reality today is, however, different. The informal housing sector provides nearly as many houses as the formal and previously public low-cost sector. While the city is struggling to provide what limited services it can, public abuse of infrastructure is rampant (Figure 5.4), indicative of a near collapse of the public planning and management system in Kitwe coupled with a lack of civic responsibility on the part of the community.

Source: Retrospect, 1962.

Figure 5.2   Kitwe City Council old coat of arms and motto

Source: Author, 1997.

Figure 5.3   Kitwe City Council new coat of arms and motto

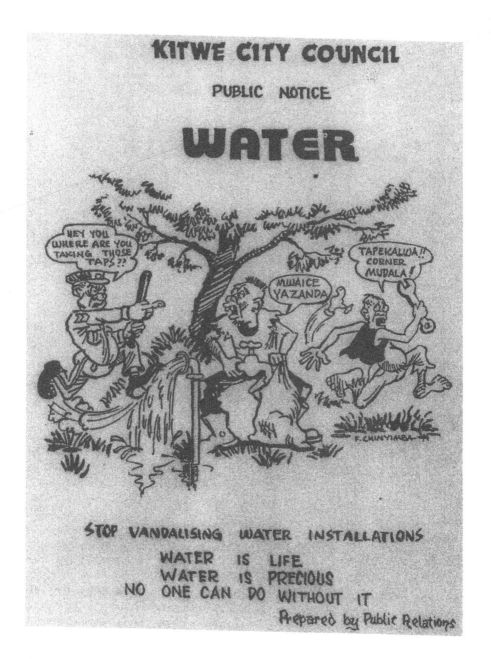

Source: Kitwe City Council Public Relations Unit.

Figure 5.4a   Kitwe City Council civic awareness poster campaign

Figure 5.4b   Kitwe City Council civic awareness poster campaign

Figure 5.4c   Kitwe City Council civic awareness poster campaign

Source: Kitwe City Council Public Relations Unit.

Figure 5.4d   Kitwe City Council civic awareness poster campaign

Source: Kitwe City Council Public Relations Unit.

Figure 5.4e   Kitwe City Council civic awareness poster campaign

Source: Kitwe City Council Public Relations Unit.

Figure 5.4f   Kitwe City Council civic awareness poster campaign

Fieldwork for this research was conducted at various times between June 1994 and December 1995. After an initial period of collation and analysis, a second visit was made to Zambia in April 1997 lasting two weeks. To date, this research represents the most comprehensive and current attempt at developing an explanatory theory contributing to the discourse on urban management in Nkana-Kitwe and Zambia. Although earlier research on Nkana-Kitwe and Zambia exists, this is mostly narrowly thematic or dated, as discussed in Chapter 1.

A Brief History of Copper Mining at Nkana and in Zambia

Zambia's Copperbelt lies in a narrow strip 120-144 kilometres long and 48-50 kilometres wide along the border with Belgian Congo (now Democratic Republic of Congo [DRC]). The country possesses six per cent of the world's proven reserves of copper ore estimated at 700-846 million short tons (Ministry of Environment and Natural Resources, 1992).

The history of copper mining in Zambia predates the advent of European prospectors. Ancient artefacts and workings point to the mining of copper for tools, ornaments and monetary purposes before the coming of Europeans (Phillipson, 1977 and Roan Consolidate Mines Ltd., 1978). In certain cases, prospectors relied on the knowledge of local people to help in 'discovering' copper ores. Robinson (1933) refers to ancient workings found at Bwana Mkubwa and Nkana and how that an 1899 expedition owed the founding of Kansanshi mine to pre-historic workings shown to them by Chief Kapiji Mpanga.

In 1889, Cecil Rhodes of the British South African Company (BSA) obtained from Paramount Chief Lewanika of the Lozi absolute power over a loosely delineated Barotseland. Claiming to be a representative of the British Crown, Rhodes offered Lewanika protection from the Matebele raiders, British residence, mineral royalties, £2000 and a gun boat in return for:

> ...the sole, absolute and exclusive and perpetual rights and powers over the whole of the territory of the Barotse nation, or any future extension thereof....to search for diamonds, gold, coal, oil, and all other precious stones, minerals, and substances (Slinn, 1972).

The legitimacy of this agreement has been discussed fully by Ndulo (1987) and Chileshe (1987), both of whom doubt its validity on grounds that deceit was involved and that the parties to the agreement did not have the same understanding of the law.

Although most of the copper deposits had been found by 1903, the lack of transport infrastructure hindered their full exploitation until the extension of the railway line from Livingstone to the Copperbelt in 1909 provided a means of transporting the copper from, and coal into, the mines. Notwithstanding the availability of a railway line, most of Zambia's copper reserves were still considered uneconomical to mine because of a copper content of between three and five per cent of the oxidized ores (compared with DRC's 15-25 per cent). Apart from two mines (Bwana Mkubwa and Kansanshi), both of which had closed down by the end of the First World War (1918), Zambian copper fields remained undeveloped until the discovery of sulphide ore bodies, which although containing only three to six per cent copper, were more valuable than oxidized ores of a similar copper content.

After a period of relative inactivity (1913 to 1923), the BSA company (which had acquired all mining rights in the territory) granted a number of companies concessionary rights to prospect for copper in the territory. It was as a result of this prospecting that it was discovered that Zambian copper ores, unlike DRC's, entered a sulphide zone after a depth of 100 feet, thus offering a regular ore formation, simpler and cheaper to extract (Bateman, cited in Davis 1933). Comparatively speaking, therefore, Zambia was realized to possess great advantages over other copper producers, among which were:

- the high grade ore averaging three to six per cent copper compared to 1.4 per cent in the USA;
- the ease with which the regular sulphide ores could be mined made Zambian mining relatively cheap; and
- cheap and efficient supply of labour.

The above advantages combined to make Zambian copper the cheapest at £23 per long ton in 1932 compared to £38 per ton for all American mines.

Functional Evolution of Nkana-Kitwe

Kitwe's development began with the discovery of rich sulphide ore deposits at Nkana by the Bwana Mkubwa Company in 1927. In addition, other deposits had been found at Roan Antelope (now Luanshya), Mufulira, Chambishi and Nchanga. Because of the great expenses involved in developing the mines at Nkana and Nchanga, three companies came together in 1931 to form the Rhodesian Anglo-American Group based in South Africa. Following the production of copper at Nkana in 1931, other shafts were sank and commissioned at Mindola and the South Ore Body (SOB) in 1935 and 1956 respectively (Figure 5.5).

Meanwhile, other mines at Bwana Mkubwa, Kansanshi, Roan, Mufulira and Chambishi were already in production or being re-developed. It was the advance in prospecting technology leading to the discovery of sulphide ores which stimulated the exploitation of Zambia's copper ore reserves. Table 5.1 is an estimate of the copper ore reserves at Nkana in 1962-63.

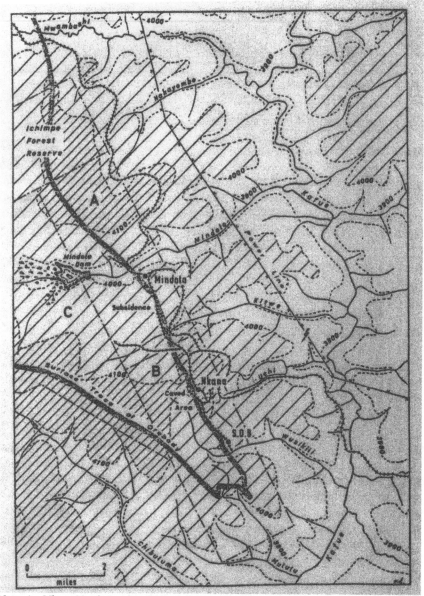

Note: A - Nkana North Land and Mineral Area; B - Nkana South Land and Mineral Area;
    C - Nkana Special Grant Area.

Source: Kay, 1967, p.134.

Figure 5.5   The site of Nkana-Kitwe

Table 5.1   Estimates of the ore reserves at Nkana in 1962 and 1963

Mine	1962 (short tons)	1963 (short tons)	1962-63 output/month (short tons)	Lifespan (Years)
Nkana	23,682,400	22,915,900	90,000	21
SOB	17,293,200	17,827,600	110,000	13
Mindola	82,838,500	83,845,600	275,000	25

Source: Adapted from Kay, 1967, p.135.

Assuming no more ore reserves and 1962-63 production, all mines at Nkana should have gone out of production by now. In fact, by 1974 national production had began to decline both as a result of diminishing reserves and high oil prices leading to high production costs (Bull and Simpson, 1993). Alluding to the future of copper mining in Zambia, the Ministry of Mines and Mineral Development states that 'developed ore resources within ZCCM properties have the economic potential to sustain copper output at current levels for at least 15 years' (Ministry of Mines and Mineral Development, 1992). There, however, exist other ore bodies which have not been developed (Konkola Deep) or mines which have temporarily closed (Chambishi), both with substantial ore reserves. In addition, with new investment, deep mining at Nkana and Mufulira could also significantly extend the lifespan of the copper industry in Zambia (Bull and Simpson, 1993). Presently, a re-privatization programme is going on with some mines already sold to private companies.

Population Growth

The development of Zambia's Copperbelt towns in the 1930s attracted a huge rural population. What was once thought as a transitory African labour force has progressively assumed a·sedentary character. The general population had risen from 30,000 in 1949 to 338,207 in 1990 (Table 5.2).

In 1970, Kitwe's public township boundary was extended to include Kalulushi, Chibuluma and Chambeshi, but the three were excised in 1977 to subsequently form Kalulushi Municipal Council. The relative population distribution within settlements is also shown for 1990 (Figure 5.6).

Table 5.2 Population of Nkana-Kitwe, 1949-50

Year	Population	Year	Population
1949	a 30,000	1975	350,000
1957	86,000	1978	260,000
1963	104,690	1980	266,286
1964	156,000	1981	314,794
1969	232,000	1982	350,000
1973	249,000	1987	b 450,000
1974	336,815	1990	338,207

Notes:
a.   African only population of Kitwe
b.   Probably includes outlying settlements and farming communities within the
     district boundary

Source: Various.

Economic Activity

The dominant economic activity in the city of Nkana-Kitwe is mining. Until the
late 1990s, Zambia Consolidated Copper Mines Limited (ZCCM), formed in 1982
as a result of incremental nationalization and re-organization of Anglo-American
Corporation and Roan Selection Trust in 1969 and 1973, operated Nkana Division.
The Division administered its section of the city called Nkana Mine Township, and
in addition ran one primary school, a secondary school and two hospitals which
were also open to non-mine employees able to pay for the facilities. Mining's
dominance in Nkana-Kitwe's economy is exemplified by the fact that of the 29
firms who responded to an economic survey questionnaire administered by this
author, nine of them reported Nkana Division or another ZCCM subsidiary as a
business partner with trade levels ranging from 2-90 per cent of their turnover. In
the early 1970s it was estimated that about 32 per cent of all employees in Kitwe
were employed by Nkana Division. Being a primary industry means that for every
one mining job there were two others dependent on it (City of Kitwe Development
Plan 1975-2000, Survey of Existing Conditions, p.84). Although the City of Kitwe
Development Plan acknowledged the possible adverse effects this over-dependence
on mining would have on Kitwe if the mine closed, this possibility did not affect
plan formulation. Prepared between 1972 and 1975, a period of moderate
economic growth, the development plan's assumption was of a continued slow
growth. The reality however, is that this growth was too small compared to
population growth and had, by 1992, reversed to a decline (Table 5.3).

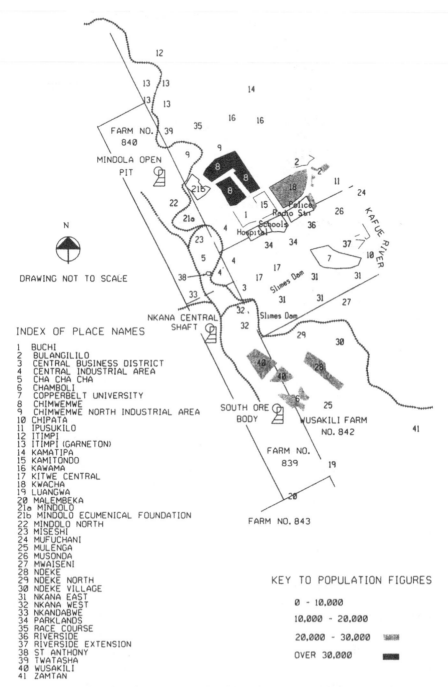

INDEX OF PLACE NAMES

1   BUCHI
2   BULANGILILO
3   CENTRAL BUSINESS DISTRICT
4   CENTRAL INDUSTRIAL AREA
5   CHA CHA CHA
6   CHAMBOLI
7   COPPERBELT UNIVERSITY
8   CHIMWEMWE
9   CHIMWEMWE NORTH INDUSTRIAL AREA
10 CHIPATA
11 IPUSUKILO
12 ITIMPI
13 ITIMPI (GARNETON)
14 KAMATIPA
15 KAMITONDO
16 KAWAMA
17 KITWE CENTRAL
18 KWACHA
19 LUANGWA
20 MALEMBEKA
21a MINDOLO
21b MINDOLO ECUMENICAL FOUNDATION
22 MINDOLO NORTH
23 MISESHI
24 MUFUCHANI
25 MULENGA
26 MUSONDA
27 MWAISENI
28 NDEKE
29 NDEKE NORTH
30 NDEKE VILLAGE
31 NKANA EAST
32 NKANA WEST
33 NKANDABWE
34 PARKLANDS
35 RACE COURSE
36 RIVERSIDE
37 RIVERSIDE EXTENSION
38 ST ANTHONY
39 TWATASHA
40 WUSAKILI
41 ZAMTAN

KEY TO POPULATION FIGURES

0 - 10,000

10,000 - 20,000

20,000 - 30,000

OVER 30,000

Source: Kitwe Water Supply Project, 1990, Table 3.3-1, drawn by author.

Figure 5.6 Population distribution within Nkana-Kitwe in 1990

Table 5.3   Population/employment ratios in Nkana-Kitwe, 1956-94

	1956	1961	1969	1975	1987	1992	1993	1994
Population (000s)	72	108	232	350	450	340	349	357
Employees (000s)	31	34	46	48	64	66	63	61
Pop/Emp Ratio	2.3	3.2	5	7.3	7	5.2	5.5	5.8

Sources: City of Kitwe Development Plan, Survey of Existing Conditions; and estimates from various CSO publications.

Not only has Nkana-Kitwe's potential tax base decreased in relation to its population, but the amount of total earnings is likely to be limited because of the diminishing growth in the formal employment sector and the low wages of the African employees who have taken over from the highly paid Europeans after independence, as shown below.

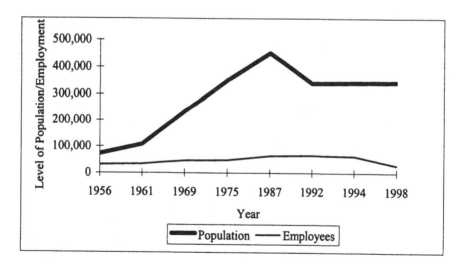

Sources: City of Kitwe Development Plan, Survey of Existing Conditions; and estimates from various CSO publications.

Figure 5.7   Relative growth of population and employment in Nkana-Kitwe

Table 5.4   Wage differentials between African and non-African workers in Zambia in 1964 and 1968

	Average Annual Earnings 1964 (K)	Average Annual Earnings 1968 (K)
All African Wages	382	789
African Miners' Wages	732	1,248
All non-African Wages	3,294	5,150
Non-African Miners' Wages	4,875	7,604

Source: Bostock and Harvey, 1972.

The ratio of all non-African to African wages at independence was 9:1, this improved to 6:1 in 1968 partly as more Africans were allowed to do skilled jobs and also due to the huge increases in wages awarded to the Africans in 1967 following the Brown Commission (Bostock and Harvey, 1972). These low incomes have had ramifications on the people's ability to pay for services and obtain access to land and housing as will be shown in the next three chapters.

The Origin of the Twin-Townships

Although copper was discovered at Nkana in 1905, the mine did not open until 1923 and production only commenced in 1932. However, by 1929, a mine township had already been established at Nkana and a management board constituted in 1935 (Copperbelt Development Plan, n.d). Notice therefore that the formal basis for 'twin' mine and public township was applied retrospectively to Nkana. The African compound consisted of grass-thatched roundavels and the European houses of grass-thatched bungalows. Figures 8.1 and 8.2 illustrate the type of houses built for the two communities. In anticipation of the growth of a non-mining population on the outskirts of Nkana mine, coupled with the mining company's desire to rid itself of the social burden of housing its own employees, the Territorial government decided to co-operate with the mining company to develop a public township adjacent to Nkana mine. For this reason, it was suggested by the Governor in a meeting with representatives of the mining company that the African mine compound be moved to another site so that it would not be situated between the public and mine township, arguing that it was not good to pass through the African compound to reach the railway station (ZNA: RC/1427: 2 May 1929).

By the end of 1929, Bwana Mkubwa Copper Mining Company which worked Nkana mine forwarded two proposed layouts of the public township to Sir Maxwell, the Governor. One was a curved street plan designed to fit the topography and the other was a straight street plan. H.S. Munroe, representing

Bwana Mkubwa, indicated that the company was ready to proceed with the layout of the township once instructed by the government (ZNA: RC/1427: 16 December 1929). In further correspondence to the government, Mr C.W. Dossett of Bwana Mkubwa writing to the Chief Secretary J.A. Northcote assured the government of his company's co-operation in the matter of the mine and government townships layout and that the company assumed that the two townships would be served by the same water and sanitary system (ZNA: RC/1427: 28 December 1929).

Despite Bwana Mkubwa's willingness to work with the government in developing the proposed public township, there was very little progress, prompting Mr. Munroe to write to Mr Northcote the Chief Secretary expressing anxiety at the lack of progress and informing the government that the company was already preparing plans for its offices and houses to be built in the government township (ZNA: RC/1427: 8 February 1930). The Town Planning Board, to which the two proposals had been forwarded, responded to the Chief Secretary's inquiry that both proposals had been turned down and a new layout was being recommended (ZNA: RC/1427: 26 February 1930).

Shortly afterwards, a meeting between the government and the mining company was called to consider a sketch layout of the proposed government township prepared by the Director of Surveys before it went to the Town Planning Board. During this meeting, Mr Munroe indicated the company's preference of plots assuming the final plan followed the sketch layout. Subject to agreement on terms, the Governor was prepared to approve the company's plot preferences and was also willing to grant special permission for the company to proceed with development prior to opening the town to the general public. The Director of Surveys also agreed to the possibility of further surveys. It was also proposed that owing to the large number of Africans, two beer halls be built. A proposal the Governor welcomed subject to the conditions that control vests in the government and that the net profits from sales would be used to develop African areas (ZNA: RC/1427: 6 March 1930). Notice that the latter condition sought to model the Durban System referred to in Chapter 4.

After due consideration of the layout as designed by the Director of Surveys, the Town Planning Board through its secretary, wrote to the Chief Secretary offering its approval on account of the urgency of the mining company's requirements. In approving the sketch layout, the Town Planning Board also proposed that land to the north of the proposed government township be reserved for future expansion. The only exception offered by the board and one which led to segregated development was that the African compound, beer hall and compound managers house sites be placed on the opposite side of the Kitwe stream, so that the distance from the residential area of the town (assumed here to be European) to the compound (assumed African area) was about one mile (ZNA: RC/1427: 10 and 20 March 1930).

Having progressed thus far and largely as a result of pressure from the mining company, the government put a hold to the development of the public township. The Chief Secretary, D.M. Kennedy writing to Mr Munroe, said that he had recommended to the Governor that work on the township did not proceed until aerial photographs were available, but that the company could proceed with the development of its allocated sites and the choice of a hotel site for its employees (ZNA: RC/1427: 10 May 1930). The other possible reason why the government might have decided to stall at this point is probably due to advice given by a Dr Alexander for the immediate appointment of an engineer experienced in town planning (ZNA: RC/1427: 8 April 1930). Shortly afterwards, Mr C.C. Reade arrived in the territory and was appointed as Town Planning Engineer and was also expected to act in an advisory capacity to the Town Planning Board (ZNA: RC/1427: 30 October 1930). Disturbed by the government's decision to put on hold the development of the public township and sensing this might result in changes to the layout plan, the mining company wrote to the Chief Secretary D.M. Kennedy stating that it contemplated an expenditure of about £20,000 on offices and residences and was seeking assurance that no changes would render less desirable the chosen sites (ZNA: RC/1427: 25 May 1930). Responding to this request, the acting Chief Secretary R.S.W. Dickinson agreed that no changes would be made to compromise the company's best site choices (ZNA: RC/1427: 28 July 1930). An assurance further reiterated by the Chief Secretary when he stated that the government Town Planning Engineer would consult with the company on the layout of the public township (ZNA: RC/1427: 21 and 24 November 1930).

In discussions with the acting Governor and acting Chief Secretary, Major Pollak the Managing Director of Rhodesia Anglo American Corporation (the successor to Bwana Mkubwa and holding company to Rhokana mining company at Nkana) requested that a hotel be sited far away from the company's chosen residential plots in the public township and also expressed his displeasure at the large numbers of Africans walking through the mine European residential area on their way to the second class (a colonial term used to classify African and Asian shopping centres) trading areas. Major Pollak suggested the establishment of a few trading sites in the African mine compound, a suggestion noted by the acting Chief Secretary (ZNA: RC/1427: 10 July 1930).

However, by 1931 there was still limited progress on the public township and on being asked by Major Pollak when this would be established, the Governor could only respond by saying that aerial surveys were being awaited after which Mr Reade's (Town Planning Engineer) advice would be sought. The Governor also promised that a proper layout of the public township would be ready before the year-end. Keen not to have its best site choices compromised by possible changes in the layout, Major Pollak suggested that the company be allowed to choose its neighbours to the north. The Chief Secretary's response was that the Town Planning Engineer would consult with the company (ZNA: RC/1427: 12 February 1931).

When the proposal for the township was finally prepared by Mr Reade, the Town Planning Engineer, it envisaged the consolidation of the existing mine township into one town. This proposal was rejected by Major Pollak who, infuriated at the length of time it was taking to establish a public township which, it was hoped, would provide opportunities for private accommodation and free the company of unproductive investment in housing, suggested that the company be allowed to develop its own township to meet the requirements of residential and trade (ZNA: RC/1427: 27 June 1931).

Rhokana's apprehension for a unified plan largely lay with the possibility that there might one day be established a popularly elected local authority (implicit acknowledgement of the illegitimacy of the colonial political and economic order) which would interfere with the present arrangement of a separate mine township which was not subject to certain provisions of ordinances governing the administration of public townships (ZNA: RC/1427: 17 August 1931). Although the government's proposal for a unified development made economic sense, this rational attempt at planning could not be pushed too hard against the wishes of Rhokana and the Acting Governor could only feebly respond by saying:

> I thoroughly understand the attitude of the mining company in regard to possible control by local authorities, which is the lion in the path of unification of development... We were exploring the possibility of getting some form of unified development without committing the mining company in any way to render themselves liable to local authorities. We have been only throwing out feelers... (ZNA: RC/1427: 17 August 1931).

One of the chief reasons for rejecting a unified development was the future possibility of heavy local government rates and taxation against both the mine plant and township leading to higher production costs and lesser profits. The attitude of the mining company was that for an industry such as theirs, it was only the supreme legislature which should value and tax it, in which case the local authority becomes insignificant. While sympathetic with this view, the Acting Governor wondered if Rhokana would be able to protect themselves from future governments. As for Mr. Reade, the author of the unified plan, when pressed by Rhokana's representative to give an undertaking, responded by saying that it was not possible to give an all time guarantee of immunity from local taxes against mining plant and townships (ZNA: RC/1427: 17 August 1931). Similar views concerning the difficulty of giving assurances to the mining company protecting them from future local authorities were expressed by the Governor, Sir Maxwell (ZNA: RC/1427: 7 September 1931) and the Chief Secretary, D.M Kennedy (ZNA: RC/1427: 5 October 1931).

A number of discussions between government officers and mining company representatives aimed at arriving at a common policy position on the public and mine township were held (ZNA: RC/1427: 17 August 1931; 24 August 1931; 26 November 1931). In 1932, the colonial governor Sir James C. Maxwell furnished the Secretary of State Sir Philip Cunliffe-Lister at the Colonial Office, details of proposed plans for establishing the two townships at Nkana arising from discussions with Rhokana Corporation. Rhokana, a subsidiary of Rhodesian Anglo-American Corporation, held land and mineral concession rights at Nkana from BSA Company in five farms of approximately ten square miles each. Three of these (Farm Numbers 839, 840 and 842) were held in fee-simple and the remaining two (Farm Numbers 841 and 843) were held on 50-year leaseholds (Figure 5.8).

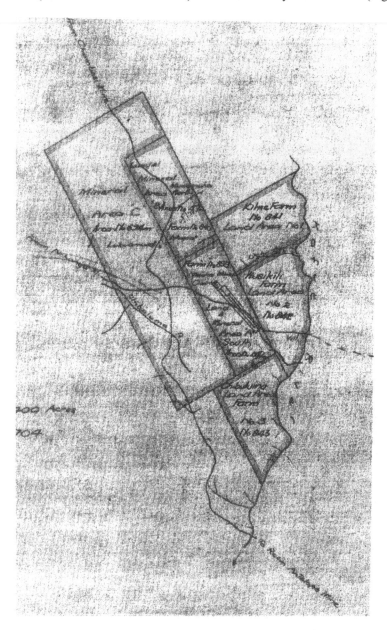

Source: PRO: CO795/50/36295.

Figure 5.8   Land held by Rhokana at Nkana-Kitwe in 1932

The following proposals for Nkana were offered by the Governor to the Colonial Office as a model for all mine areas on the Copperbelt (PRO: CO/795/50/36295: dated 7 January 1932).

Mining township: It was proposed that the area of Farm 841 'Kitwe', lying west of the railway (Figure 5.9) held on a 50 year lease by Rhokana Corporation, should be excised from the parent parcel and conveyed to Rhokana Corporation in fee-simple to form part of the mining township and be developed solely for mining and housing for mine employees. Although the government would endeavour not to interfere with the development of the township, it would reserve the right to prevent the introduction of trade into the area by refusing to grant licences.

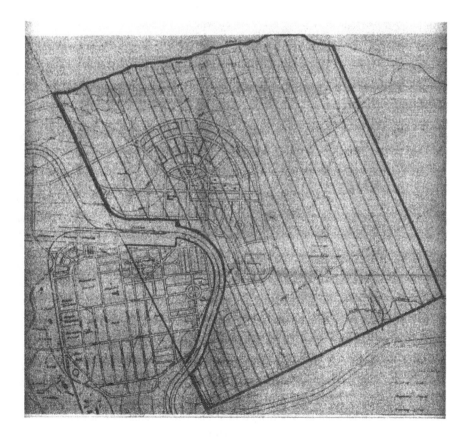

Source: PRO: CO/795/50/36295.

Figure 5.9   1932 Version of the proposed mine and public townships at Nkana

Public township: Concerning the public township the following were proposed:

(i) the corporation or a subsidiary to undertake to develop and provide services over a piece of land held on a 999 year lease granted by the government and the former would constitute the local authority under s5 of the Township Ordinance. The establishment of any other local authority apart from the corporation would only be possible if the local authority agreed to take over the public services at an agreed basis;

(ii) in addition to making profits from dealing in the land, the governor would consider granting the corporation half the revenue from trade licences and half the fines collected in the public township;

(iii) it was further proposed that the railways company would make their own arrangements with the corporation regarding land and that no other township would be created by the government within a specified radius of Nkana so long as a considerable portion of this public township remained undeveloped; and

(iv) lease conditions were to provide the government power to demand the re-conveyance of land necessary for present and anticipated use, with a provision for compensating the corporation on present use value of the land.

The above proposals were forwarded to the Colonial Office with the reasons that, if adopted, they would provide the quickest way of establishing the public township at Nkana using the financial stability of the corporation and making use (by way of extension) of existing service lines in mine area. The proposals, if accepted, would also relieve the Territorial government of a heavy financial burden.

While the Territorial government was more sympathetic with Rhokana's wish to have absolute control of their township, the Colonial Office treated the matter with grave concern, questioning the morality of handing over control of its people to a foreign profit-making company for financial reasons, 'it is not our business to increase the profits of a largely American absentee landlord by facilitating their control over our own people' (PRO: CO/795/50/36295: dated 9 September 1932). In addition, the Colonial Office was concerned about potential problems should it be necessary for the government to take over control of the township. In the face of the growing need for a public township at Nkana and yet given the unfavourable financial situation of the Territorial government, the Colonial Office agreed to the proposals subject to the following modifications:

(i) the public township to be established by Rhokana on land leased for 99 years or preferably 50 years with a provision for compensation in respect of improvements at the end of the lease; and

(ii) Rhokana to appoint the majority on the Township Management Board and the government to be represented by its nominated member.

The Colonial Office expected this arrangement to break down even before the expiry of a 50-year lease.

The Territorial government responded strongly to the Colonial Office's modification of the proposals, arguing on behalf of Rhokana that the heavy capital outlay in developing the public township justified a longer lease than the counter proposal of 99 years or 50 years. Against the feared excesses of control, it was argued that all bye-laws made by the Board would be subject to the approval of the governor-in-council. The prospect of such abuses, the Territorial government argued, would be further curtailed in that the enforcement of sanctions as a result of the violation of bye-laws would be before a government judicial officer. The Territorial government once again reiterated its inability to incur such an expense in the development of a public township and insisted that Rhokana should be allowed generous terms of tenure. This did not convince the Colonial Office, who directed that Rhokana should be granted a 99-year lease in respect of Subdivision B of Farm number 841 Kitwe, to replace the existing 50-year lease for the development of a public township.

This turn of events infuriated Rhokana, who declined the offer advancing the argument that the failure to attract developers at Luanshya could be attributed to government's refusal to grant freeholds in preference for 99-year leases. In a strongly worded statement to the Chief Secretary, Major Pollak observed:

> In these circumstances for the government now to offer the corporation the same period of lease as that which any individual can obtain in respect of a residential or business plot is, I submit with all respect, not at all reasonable. If, as I contend, the poorness of the sales effected by government at Ndola and Luanshya was due to restricting the leaseholds to 99 years, what hope would the corporation have at Nkana if it could not even grant a 99 year lease? (PRO: CO/795/62/5587: dated 23 August 1933).

Furthermore, Rhokana argued that Nkana had since lost its prominence to Ndola and that a public township at Nkana was now no longer necessary. Based on this observation, Rhokana suggested that present arrangements of tolerating trade in the mine area should continue except that Rhokana should be allowed to erect reasonable structures for rent to the traders, being unwilling to part with any land or grant long leases to land in its mine area. The condition attached to this proposal by Rhokana was that the government should undertake to allow trade to continue in the mine area notwithstanding the possible development of a public township. The Colonial Office's refusal to accept the grant to Rhokana of a 9-year lease diminished the prospect of a public township at Nkana for some time and the Territorial government could only agree with Rhokana's proposal to allow trade to temporarily continue in the mine area. The Colonial Office too agreed with allowing trade to continue in the mine area, but advised the Territorial government to keep open the possibility of developing a public township. Still concerned about keeping the powers of Rhokana in check, the Colonial Office advised the Territorial government to aim for the following conditions:

(i)   Rhokana to apply for the declaration of Nkana as a mine township - a situation which, under the Mine Townships Ordinance would prohibit trade in the mine township, except with the permission of the Governor;

(ii)   no trading within the mine area without the approval of the Governor - a condition which could be used should Rhokana not want to apply for the declaration of Nkana as a mine township or before the area is declared a mine township;

(iii) limit the number of shops to the existing level, any new shops to seek the permission of the Governor;

(iv)   undertake to allow trading for ten to 15 years irrespective of the development of a public township; and,

(v)   in return for allowing trading activities within the mine area, Rhokana to let houses to government at reasonable rents.

Having settled on the less favourable and temporary solution of allowing Rhokana to build shops within the mine area for rent to the business community, the next sticking points were on the length of the leases to be granted by Rhokana to shop owners and the amount of rent to be charged. Whereas the government favoured a much longer lease, or shorter ones but with renewal clauses, Rhokana on the other hand preferred monthly tenancies or one year leases. The government counselled Rhokana that the assurance not to establish another township outside the mine area during the agreed period might well depend on agreement being reached on the rent and tenurial questions. Rhokana were incensed by what they perceived as a deprivation of use rights especially that the land was held in freehold with no conditions attached specifying the use to which this land could be put. The Colonial Office took strong exception to Rhokana's objections to give security of tenure to traders within Nkana mine area, an objection which the Colonial Office interpreted as aimed at getting the kind of control which the government did not wish them to have. This observation was explicitly accepted by Rhokana, who said that they would be justified to eject any trader they found supporting industrial action against the corporation. With the death of Major Pollak of Anglo-American Corporation in 1934 (a major player in the negotiations with Rhokana), the government was looking forward to making progress in its negotiations with Rhokana.

After protracted negotiations between the Territorial government and Rhokana, agreement was finally reached on a number of points, including the following:

(i)   Rhokana was to apply for Nkana to be declared a mine township;

(ii)   trading was to be limited to the number of existing shops, any further applications could only be approved by the Governor; and,

(iii) the government was to allow trade to continue in the mine township for 20 years after the agreement.

Financial constraints could not allow Rhokana to construct trading premises for rent in the mine area and a suggestion that traders be asked to put the money upfront for the construction of these premises was seized upon by Sir Hubert Young the Governor, who reasoned that 'If traders were to put up the money, it re-opened the possibility of the traders' stores being erected in a public township and not trading in the mining township' (PRO: CO/795/76/25628: dated 27 September 1934).

Following Rhokana's failure to erect reasonable trading premises, the Territorial government approved the continuation of temporary trading arrangements while plans were underway for the establishment of a modified form of public township. Meanwhile, Rhokana applied for, and the government under Notice 2 of 1935, declared Nkana a mine township under the Mine Township Ordinance. For all intents and purposes, Nkana could be classified as a company town, defined by Allen as:

> ...a community which has been built wholly to support the operations of a single company, in which all homes, buildings, and other real-estate property are owned by that company having been acquired or erected specifically for the benefit of its employees, and in which the company provides most public services (Allen, 1966, p.4).

As to the private nature of Nkana Mine Township, Rhokana maintained that streets in the mine townships were private and therefore the company reserved the right to close them to the public at any time. It was only after government intervention that a compromise was reached, resulting in an amendment to the mine townships ordinance in 1939 (PRO: CO/795/110/45228: dated 13 June 1939). This amendment provided for the declaration of streets in the residential areas (European section) as public and the mining plant and streets in the compound (African section) as private, possibly to facilitate the control of African movement in compounds. Because it was a private township, Nkana was exempted from a number of provisions relating to the administration of public township, for example, when general property rates were eventually introduced in the public township, these were not levied in Nkana.

In April 1935, the Territorial government was pursuing further negotiations with Rhokana in which it was proposed that the corporation alienates and leases out a limited number of stands, provides the infrastructure and that the new public township be managed by the Mine Township Board to include an appointed government officer. In return, the corporation was to be reimbursed its expenses on the provision and maintenance of services from the premiums and rentals of the alienated stands. These proposals were forwarded by the Territorial government to the Colonial Office in August 1935 with the hope that they would be accepted with few changes. However, when the final communiqué was published, it differed markedly from the above proposals. The following were concluded as comprising the procedure to be followed in the establishment of a public township at Nkana (PRO: CO/795/76/45077: dated 17 August 1935).

(i)   Stands would be alienated and allocated by the government for 99 years on land acquired from Rhokana Corporation Ltd.

(ii)   Stands for trade purposes were to be limited to the existing numbers with a possible additional two stands in each trade sector with preference in allocation being given to existing traders - this restriction being effective for 20 years from 1 April 1936 was designed to reward existing traders who were to advance the money for the establishment of the public township.

(iii) Advance payments for the stands were calculated to allow for the provision of social and physical infrastructure and subsequent income from the sale of stands and rentals were to be used for the maintenance and extension of the town. Thus was to be built a public township at no expense or profit to the Territorial government.

(iv)   In addition to limiting the number of traders in the public township, the government undertook not to establish another public township within a radius of 16 Km (ten miles) of Nkana's smelter stack for a period of 20 years from 1 April 1936 unless population or other pressures should warrant doing so.

(v)   In a gesture of goodwill, Rhokana agreed to:

       (a)   give up the land required for the public township without compensation; and,

       (b)   to carry out engineering works and to provide the services in the public township at cost price.

(vi)   Local residents appointed by the Governor would constitute the Management Board under the chairmanship of the District Officer.

The public township was established under General Notice 397 of 1935 published on 12 September 1935. This notice will be referred to at considerable length in the next chapter when discussing the financial structure before the introduction of general property rates in Kitwe.

With these firm proposals, it was expected that the layout of the new township would commence in 1936 and that trading was to cease in the mine township on 30 September 1937. The limitation on the number of traders within the public township and the undertaking not to allow the development of another public township outside it effectively meant that the public township, like the biblical Jericho, was shut in - no trader went in and none went out – a condition which earned it the title of a 'closed township' (Kay, 1967; City of Kitwe Development Plan 1975-2000; Kitwe Street Plan). The township was called Kitwe after the farm on which it was laid out. There is no doubt therefore, that although Kitwe's development was made possible by an initial capital outlay from private traders, the traders and by extension the initial development of Kitwe were dependent on the mining community who at this time represented a reasonable concentration of people which, it was envisaged, would support another reasonable non-mining community outside of its own Nkana mine township.

The mining company therefore was perceived by the Territorial government as a necessary economic ally in the development of a viable public township. This strong economic position might have influenced the attitude of the Territorial government shown in the privilege enjoyed by the mining company by way of exemptions from ordinances governing public townships as provided in the provisions of the Mine Township Ordinance, and exemption from charges on land acquired in the public township by the mining company (Notice 397 of 1935). Other possible pointers to the influential position of the mining company are the indemnity which was granted to Rhokana on 12 March 1934 (PRO: CO/795/68/25528: dated 12 March 1934) protecting the company from being prosecuted for air pollution within a radius of ten miles from an identified point within the mining area and the flexibility with which the local authority dealt with rates due on mine property, a subject discussed in the next chapter. Kitwe public township was officially opened in November 1937 by the Governor, Sir Hubert Young (*Bulawayo Chronicle*, 26 September 1947).

Physical Development of the Twin-Township

In order to facilitate the discussion of the physical development of the city of Nkana-Kitwe, a chronological chart of ten year periods is used (Table 5.5). A number of factors affected the choice of these decade years. Although Nkana township had already been established by 1929, it was not until 1935 that a township management board was constituted. This forms the starting point in tracing the physical development. The arrangement of these periods in decades was also influenced by the desire to draw a chart which would nearly coincide with key dates and periods in the development of the city. For example, the 20 year closed township period from 1936-56, creation of the Municipal Council in 1954, Federal Government period from 1953-63, independence in 1964, First National Development Plan (1966-70) and so on.

Notice too that the development of the town in this section is based on date of original survey of land parcels and might therefore not coincide with actual date of construction. For example, original surveys of land in Nkana East were all done in the period 1956-65 but actual construction developed in phases spanning a period longer than the decade 1956-65 and is still being developed today.

The core of Nkana Mine Township (now Nkana West) comprising central administration, hospital, recreation facilities and European houses was laid out in a grid iron pattern and had, by 1935, assumed its present character (Figure 5.10). By 1940, Wusakili was laid out south of Nkana West as an African housing area (Tipple, 1978). To house African workers on Mindola mine four miles north of Nkana which started production in 1935, Mindolo suburb was laid out in three phases 1934-47; 1954-56; 1966-79. In what can be construed as a shift of activity from the two southern mines, this initial development of Mindolo was followed by another three suburbs - Miseshi, Cha-Cha-Cha and Mindolo North. Apart from

Nkana West, all other mine housing areas were either low- or medium-cost and provided for African workers. Meanwhile, European mine workers' housing was expanding eastwards of Nkana crossing the Ndola-Kitwe trunk road, the Luanshya-Mufulira powerline and reaching the Kafue river in 1965. This eastern wing of Nkana is presently called Nkana East (Figure 5.10).

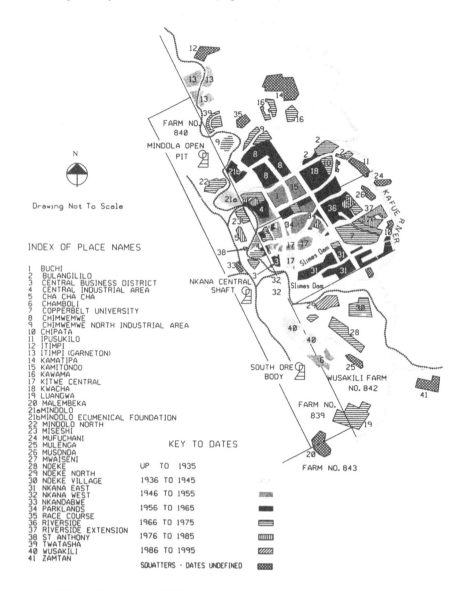

INDEX OF PLACE NAMES

1 BUCHI
2 BULANGILILO
3 CENTRAL BUSINESS DISTRICT
4 CENTRAL INDUSTRIAL AREA
5 CHA CHA CHA
6 CHAMBOLI
7 COPPERBELT UNIVERSITY
8 CHIMWEMWE
9 CHIMWEMWE NORTH INDUSTRIAL AREA
10 CHIPATA
11 IPUSUKILO
12 ITIMPI
13 ITIMPI (GARNETON)
14 KAMATIPA
15 KAMITONDO
16 KAWAMA
17 KITWE CENTRAL
18 KWACHA
19 LUANGWA
20 MALEMBEKA
21a MINDOLO
21b MINDOLO ECUMENICAL FOUNDATION
22 MINDOLO NORTH
23 MISESHI
24 MUFUCHANI
25 MULENGA
26 MUSONDA
27 MWAISENI
28 NDEKE
29 NDEKE NORTH
30 NDEKE VILLAGE
31 NKANA EAST
32 NKANA WEST
33 NKANDABWE
34 PARKLANDS
35 RACE COURSE
36 RIVERSIDE
37 RIVERSIDE EXTENSION
38 ST ANTHONY
39 TWATASHA
40 WUSAKILI
41 ZAMTAN

KEY TO DATES

UP TO 1935

1936 TO 1945

1946 TO 1955

1956 TO 1965

1966 TO 1975

1976 TO 1985

1986 TO 1995

SQUATTERS - DATES UNDEFINED

Source: Kitwe Land Register, 1995, drawn by author.

Figure 5.10   Chronological development of Nkana-Kitwe

Following agreement on the methods of establishing a public township, an estimated 180 stands (Table 5.5) were demarcated between 1936 and 1945, consisting of the European trading area (also known as first class - in reference to the social class accorded to Europeans), a few plots in the African trading area west of the former (also known as second class trading area) and part of Kitwe central residential area. The racially segregated trading zones were separated by a swampy marsh.

Table 5.5   Land parcels by date of original survey

	up to 1935	1936-1945	1946-1955	1956-1965	1966-1975	1976-1985	1986-1995	Totals
Stands	2	180	1,330	2,915	1,245	928	741	7,341
Lots	31	1	0	40	16	14	4	106
Farms	11	17	18	39	79	41	193	398
Totals	44	198	1,348	2,994	1,340	983	938	7,845

Source: Compiled by author from Kitwe Land Register, Ministry of Lands, Lusaka, 1995.

The following decade, 1946 to 1955, was a period of rapid growth. About 1,330 stands, seven times more than the previous decade, were laid out. In 1948, industrial development was allowed for the first time in Kitwe (Anon, 'General Information on the Town'). It was in this decade that Garneton (now called Itimpi), Kitwe's dormitory suburb, was planned about nine kilometres north of Kitwe off the Kitwe-Chingola road. By 1955, Kitwe central and Parklands, both European residential areas, had already assumed their present characteristics, the two being separated by the Kitwe stream. This growth in Parklands and Kitwe central could have been the result of the increase in the light and heavy industrial activity exemplified by the establishment of a large area in 1948 west of Parklands to support mining activities. This period also saw the construction of Buchi and Kamitondo African suburbs to house African non-mine employees (Tipple, 1981). At about the same time, the African Towns idea originally mooted by the Governor to the Native Industrial Labour Advisory Board in 1935 (Northern Rhodesia, 1935, cited in Tipple, 1981) and formally established in 1944 (Tipple, 1981) saw the development of Chibuluma African township situated about six kilometres west of Kitwe, much to the delight of white miners and settlers who wanted a ten kilometer exclusion zone to preserve the white tone of Nkana-Kitwe. Where African settlement was established within proximity of white areas, it was mainly out of labour needs and might also reflect a change in colonial attitude, as the old conservative order was being replaced by a new university trained liberal cadre of young men and women but, even then, residential areas were spatially segregated using planning tools and labour laws. Given this unprecedented growth in Kitwe's secondary industry, it was necessary to provide additional trade and financial services in the 'first class' town centre. The undertaking to limit the number of stands in the town centre was waived and additional stands were laid out in 1953 (Kay, 1967) at the edge of the town centre and appropriately called Coronation Square (later changed to Kaunda Square, after the first Zambian President). The town attained municipal status in 1953 with the Management Board becoming a Municipal Board and then Municipal Council in 1953 and 1954 respectively.

The expiry on 1 April 1956 of the 20-year agreement made between the Territorial government and the business community in Kitwe ushered in a decade of increased activity in the development in Kitwe. The period from 1956 to 1965 still remains the most outstanding in Kitwe's growth. The relative boom of this decade is probably attributable to the corresponding economic boom as copper production more than doubled from 276,000 long tons in 1950 to 579,000 in 1960 (Turok, 1989). An estimated 2,915 stands, more than twice the number laid in the last decade, were demarcated (Table 5.5). A new European residential area called Riverside was opened up east of Parklands but separated from it by the Luanshya-Mufulira high voltage powerline. Notice that until now, except for Buchi and Kamitondo, public African housing had not been given real attention.

The decade 1956-65 saw the establishment of a further two African settlements north of and outside the Kitwe Farm boundary. These are Kwacha and Chimwemwe. By 1965, a segregated town had developed in which the two planning authorities (mines and municipal council) had each planned and developed separate areas for Europeans and Africans. With reference to Figure 5.11, Kitwe's public African suburbs north of Line XX were separated from the European suburbs called Riverside, Parklands and Kitwe central in the south by a chain of institutions (hospital, schools, sports ground, transmitter station and police camp) forming a buffer zone. Similarly, other African housing south of Line YY was separated from European housing by a swampy marsh and railway line. Thus we see racial prejudice translated into physical space. Kitwe's light and heavy industry kept on growing, with additional plots demarcated adjacent to the existing ones west of the Kitwe-Chingola trunk road.

In May 1966, Nkana-Kitwe was declared Zambia's second city after the capital city Lusaka. Although the Territorial government had in 1935 declared its intention to establish Ndola as a district centre, it seems that Kitwe grew beyond the government's expectations.

> It is the policy of the government to develop Ndola as the commercial and distributive centre of the Copperbelt and to provide only retail trading facilities elsewhere... (PRO: CO/795/76/45077: no date).

Having lost its prominence to Ndola for a while, a situation which led Major Pollak of Anglo-American Corporation to comment that the establishment of a public township at Nkana was no longer warranted (PRO: CO/795/62/5587: dated 23 August 1933), Kitwe came from behind to establish itself as the second city in Zambia by 1966.

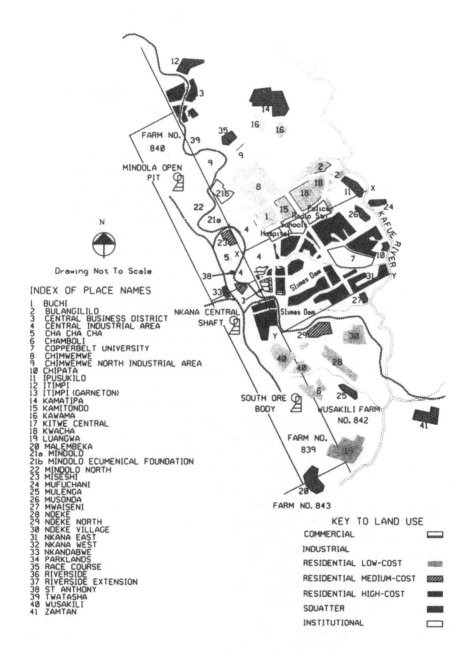

Source: Author.

Figure 5.11 Distribution of land uses in Nkana-Kitwe, 1999

The decade 1966 to 1975 was of particular importance in that it presented the new Zambian government with two opportunities to chart out its course in settlement policy. Two development plans were promulgated and implemented between 1966 and 1976. In their analysis of settlement policy in the First National Development Plan (FNDP, 1966-70), Kasongo and Tipple state that the FNDP:

> ...reacting to the almost exclusive rental market of the colonial period, encouraged home ownership. Mindful of the high cost of building complete houses, the government proposed site and service schemes at a time when these were a real novelty (Kasongo and Tipple, 1990, p.151).

The period 1966 to 1975 saw the establishment and growth of site-and-service schemes at Bulangililo, Luangwa, Kawama and Twatasha (Figure 5.10) demarcated and serviced to allow for the transfer of people from unauthorized settlements which, by 1975, had grown to more than ten major ones (Figure 5.10). In addition to encouraging home ownership in the low income sector, this period saw also the establishment of a medium-cost home ownership scheme at Ndeke and the beginnings of high-cost home ownership in Riverside extension. By 1975, Kitwe's existing industrial site had reached its extreme western limit and a new area was opened north of Chimwemwe. A notable institutional development during this decade is the Zambia Institute of Technology (now Copperbelt University) which adjoins Riverside on the north, the Luanshya-Mufulira powerline on the west and abuts the Kitwe stream in the south. In terms of territory, Kitwe city boundaries were extended in 1970 to include Kalulushi, Chibuluma and Itimpi enclosing an area of about 1,800 square kilometres (City of Kitwe Development Plan, p.11).

The decade 1976 to 1985 saw the continuation of a decline in the development of the formal city which had started in 1966-75, the period following the peak of 1956-65 (Table 5.5). Apart from the laying out of two high-cost areas in Riverside and Riverside extension and the filling in of gaps in the industrial area, not much else was planned except for a small area in Kwacha and the reclamation of a low lying marshy open space for a shopping complex in Parklands. Another Territorial adjustment was done in 1980 by the excision of Kalulushi, Chibuluma and Chambeshi from Kitwe and their incorporation into Kalulushi District. The remaining townships of Kitwe, Nkana and the dormitory suburb of Itimpi formed Kitwe District (Kasongo and Tipple, 1990).

It is difficult to identify any concentration of urban development in the last decade (1986-95). The fall in the number of planned plots continued but perhaps as a pointer to the diversification which must follow the decline in the importance of copper as Nkana-Kitwe's major economic activity, a record number of farms (estimated at 193, Table 5.5) were surveyed, presumably for the retrenched labour force from the mines embarked on in the early 1990s.

There is no comprehensive written history of the evolution of squatter settlements in Nkana-Kitwe. Apart from a general reference to 200 married men and 200 single men living in squatter settlements around Kitwe (Northern

Rhodesia, 1944, cited by Tipple, 1978 and 1981), most of what is available is verbal communication. The earliest identified and still existing squatter settlement is Mufuchani. This settlement was established by a man called ShiMwila (Father to Mwila) around 1952 (Pakeni, 1973), who took advantage of the Kafue River and the area's proximity to the township of Kitwe to sustain his charcoal burning business (KCC, 1976). Strict controls in the colonial days limited further developments of squatter settlements. Most unauthorized settlements therefore are a post-independence phenomenon fuelled by the removal of movement control measures and limited housing provision. For example, Itimpi, Kamatipa, Ipusukilo, ZamTan - all generated by their nearness to sources of employment, were all established between 1966-75.

Evolution of Urban Administration

This section is an attempt to relate the forms and functions of urban administration at Nkana-Kitwe to changes in the political economy of the country. With the declaration of a mine township at Nkana in 1935, a Mine Township Board of Management (MTBM) was set up with duties to ensure public order, safety and health. Although the MTBM had similar powers and responsibilities as those of the boards established for the administration of public townships, the main differences were that unlike public township boards, which were taxing authorities and accountable to an electorate, the MTBM did not collect rates or personal levies and had no allegiance to a political constituency, its officers being nominated by the mining company and not elected by residents (Greenwood and Howell, 1980).

The public township of Kitwe was, until December 1952, administered by the Kitwe Management Board (KMB) appointed by the Governor. The board comprised eight nominated members of whom one was the District Commissioner, who was also the chairman, and two representatives from the mining company (Northern Rhodesia, 1950). One of the people who served on this board and had gone on to be the mayor of Kitwe from 1960-63 was W.M. Comrie, the TUC representative mentioned in Chapter 4, who had been sent out by the British government to assist in the formation of African trade unions on the Copperbelt and had decided to stay on (Makasa, 1981; KCC diary notes for 1997). In January 1953, the Management Board was elevated to a Municipal Board and in 1954, the Municipal Board became a Municipal Council under the Municipal Corporations Ordinance of 1927 (as amended in 1929). The Ordinance provided for Councils where a majority were elected by European rate-payers and including three government officers appointed by the Governor with powers to make bye-laws and the control of African locations within their territorial jurisdiction. Although as from 1949, Africans in African Townships began to pay rates (Conyngham, 1951), those living in municipalities at the pleasure of the Europeans were still exempt from paying rates. More will be said about the financial structure for financing public services in the next chapter.

Shortly after independence (1964), the Municipal Corporations Ordinance and the Township Ordinances were repealed and, in their place, the 1965 Local Government Act was enacted to provide for the management of urban population centres while Nkana continued to be administered by the Mine Township Ordinance. The government's intention to decentralize central government by the appointment of District Governors to direct all development efforts in the Districts in 1969 led to conflicts, erosion of local authority autonomy and increasing interference from the ruling political party UNIP. This led to the enactment of the 1980 Local Administration Act which replaced the 1965 Act with the aim of 'providing for an integrated local administration system' (Rakodi, 1988b, p.39). Following these changes, introduced by the 1980 Act, Kitwe City Council became Kitwe District Council. Although the three proclaimed objectives of the 1980 Act were to decentralize power, the creation of District Councils as catalysts in the development process and to integrate the roles of the party and local government so as to remove the conflict that developed between the two before 1980 (Mukwena, 1992), real power had not devolved from the centre (Mutale, 1993). It is the dominance of the party in local government and other state organizations which, after the re-introduction of multi-party democracy in Zambia resulted in calls for the delinking of the party from government. In 1991, a new Local Administration Act was passed replacing the 1980 Act and providing for the reversion to the elected Mayor rather than appointed District Governor as head of the council. Once again, Kitwe became a City Council.

Form and Function of Urban Administration

The primary motive in the establishment of Nkana Mine Township was to provide houses and services for the mine workers free from government interference, and through this exercise, unfettered control over the work force. Control was achieved by tying the provision of subsidized shelter and services to employment and generally promoting a spirit of dependence. As a result of this, it was possible and necessary to have a township management board appointed by the mining company and hence not accountable to the residents in the mine township. Where consultation was made with the people's representatives from the township, this was merely to sound the people's sentiments, check omissions or excesses and generally to lend credibility to an otherwise undemocratic institution. The management board was not legally obliged to heed any of the advice arising from such consultations.

If the mine township's local authority was an imposition on the residents from the mining company, Africans in the public township of Kitwe (at least until 1956) found themselves under a local authority constituted from Europeans, for whom the town was built and in which the African lived at the former's pleasure. As at 1949, there was no African on the Kitwe Management Board (Clay, 1949). By 1956, the only African presence on the Municipal Council was from the African Housing Area Boards and the African Affairs Advisory Committee each of which appointed two Africans to sit, not on the full Council, but rather as observers with no voting powers on the Council's Non-European Affairs Committee - the very Committee which would debate, vote and recommend to the Council, decisions affecting the Africans in Kitwe (Northern Rhodesia, 1957a)! Even on the African Affairs Advisory Committee, the African presence was restricted to a token two Africans against seven Europeans. Rather than being equals on the Advisory Committee, the African presence was merely tolerated and was sometimes a source of complaints by some European staff who thought it demeaning that their salaries and conditions of service should be discussed in the presence of and voted upon by Africans (Clay, 1949). This attitude was prevalent in the thinking of the European - either as a colonial public officer or a private person. In Cecil Rhodes' own words, 'the Native must be treated as a child and denied the vote' (Rhodes 2, BBC Film). For the above reasons, it was only after the change in the electoral law, granting equal voting rights in 1963, that Kitwe had its first Zambian Mayor (Mr. J.P.S. Kalyati, 1964-65) and but later still the first Zambian Town Clerk (Mr. R.W. Musonda, 1972-76). While the enfranchizement of Africans in 1963 was welcome, it was then that local government became politicized, compromising the relative independence of councillors before 1963. For example, Kitwe's mayor was recently threatened by the district leadership of the ruling MMD for refusing to allow council workers to be used in selling MMD calendars, a reprimand reminiscent of the UNIP days and the conflict that existed between the council and party functionaries. In another case of party interference in urban administration, residents interviewed after 20 houses collapsed following the flooding of the Kafue River, claimed that they had been allocated plots by the Ward Councillor (*Zambia Daily Mail*, 18 March 1998).

As of 1999, Kitwe City Council had seven departments each headed by a director, thus: Administration; Engineering Services; Finance; Housing and Social Services; Legal Services; Public Health; and Water and Engineering Services. Each of the directors reports to the Town Clerk who is the Chief Executive.

The council's policy-making body is headed by the Mayor sitting with up to 25 Ward Councillors and five members of Parliament. The council also has seven standing committees namely: Establishment; Finance and General Purposes; Housing and Social Services; Licensing; Plans, Works and Development; Public Health; and Water and Sewerage Services.

Having explored the negotiations which established the twin-townships of Nkana-Kitwe and the resulting urban form and organization, the next section draws together the explanatory theory on the complex political and economic forces which have influenced urban development processes and urban management at Nkana-Kitwe.

Analysis of Power Structures in the Urban Development Process at Nkana-Kitwe

Pre-Colonial

Three institutions can be identified in the development of Nkana mine township (Figure 5.12), being Rhokana Mining Corporation, the Territorial government and the Colonial Office. Each of these three groups pursued objectives which varied with one or coincided with another. For example, it is possible to identify a strong relationship between Rhokana and the Territorial government at certain times, but not at others. Such affinities were also influenced by personal relationships in these organizations so that, as personnel changed, so did some of these relationships.

Rhokana's primary objective in the development of Nkana was largely economic. In order to exploit the rich copper ores in a remote part of the country, the mining company had to provide housing for its workers, recreation and medical care, and other services necessary in a residential area. The combined effect of the nature of mining operations and the company's desire to maximize profits resulted in a highly controlled regime bordering on military discipline, evidenced by the form of its early housing policy (barrack type) and the existence of a mine police force. In order to achieve this objective, Rhokana sought and established alliances with the Territorial government which in turn represented their case to the Colonial Office. The company was strong enough to exert its influence in decisions which affected their operations.

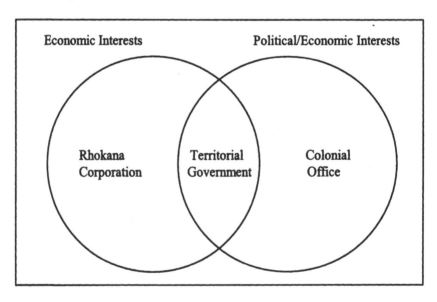

Source: Author.

Figure 5.12 Power bases in the evolution of Nkana

The Territorial government was vested with powers from the Crown to administer the country by the maintenance of law and order and thus facilitate the uninterrupted exploitation of resources. The Territorial government pursued the twin objectives of political control and economic development. Given the dominance of private companies in mineral exploitation, the government's economic objectives were limited to the implementation of fiscal policy, while on the political front it facilitated the continuous supply of labour and did the bare minimum to provide public services and to protect the native population. In return, the Territorial government levied taxes on companies and individuals.

The Colonial Office, on behalf of the Crown, decided the broad policy framework for the administration of the colonies. Although the legislative council in the territory had powers to legislate, this was subject to the approval of the Secretary of State at the Colonial Office. The Colonial Office therefore held the ultimate veto power over what went on in the territory.

Although Tipple (1978) identifies the Colonial Office in Britain, the Territorial government, mining companies, missionary societies and South African settlers as wielding political power in Northern Rhodesia, it was mainly the first three institutions which contributed to the growth of Nkana. The trading sector of the settler community, having developed into a critical mass and grown to be financially sound, did participate in the development of the public township at Nkana. Although missionary activity was present, the religious institutions are notable for their silence in affecting the development of the physical form of Nkana-Kitwe. Their only interest in the Copperbelt as a whole was their concern for the changing social and economic values as a result of rapid industrialization and missionary activity and the effect these were having on indigenous structures (Davis, 1933). The failure by the missionaries to influence the growth of the physical environment within which their adherents were expected to live fulfilled lives is regrettable, as has also been noted by Davis (1933, p.344), in that 'there is an intrinsic lack in a Christianity that widens horizons, creates new visions and stimulates desires, but fails to make its influence effective in assisting men to achieve these ideals'. Three power bases can be identified in the growth of Kitwe. These are the small time trading community (but big enough to provide capital for the development of the town), the giant mining company and the Territorial government (Figure 5.13).

Absent from this group is the active participation of the Colonial Office, notable for its strong presence in the development of Nkana. The reason being that what was about to take place at Nkana was novel and a departure from the norm and therefore the Colonial Office was wary about what was almost perceived as an abdication of duty to a private company. The same cannot be said of the presence of the small trading community which was only providing advance funding in return for a measure of monopolistic gain and was not interested in the day-to-day management of the town.

We see that the presence of an alternative capital source for the development of Kitwe somewhat softened Rhokana's stand. In order to remain a good corporate citizen and possibly to keep open the opportunities for influencing policy related to their mining operations, the mining company was prepared to surrender its lease on Kitwe farm and put its technical services to use in the development of Kitwe. After the establishment of the public township at Kitwe, the day-to-day management was vested in the local management board which was made up of eminent citizens chosen from among the electorate and public officials - a board which might have included prominent business people from the settler community and public officers.

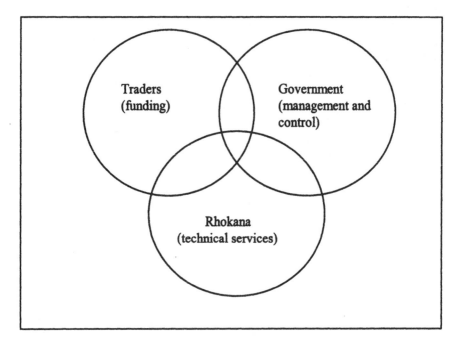

Source: Author.

Figure 5.13   Power bases in the evolution of Kitwe

Post-Colonial

Following independence in 1964, Kasongo and Tipple (1990) identify three actors controlling the growth of Kitwe - these being the mining company, Kitwe City Council and the ruling United National Independence Party (UNIP).

To this group must be added squatters, whose influence has sometimes been weak and at other times strong, depending on the prevailing political climate. For example, Kasongo and Tipple (1990) cite a case in which Kitwe's settlement plans were substantially changed by squatter action. Absent in this group is the formal trading community which had played a pivotal role in the establishment of the public township. The activities of UNIP in influencing the growth of Kitwe being mainly restricted, although not exclusively, to the informal settlements. Tipple (1976b) and Kasongo and Tipple (1990) observe that UNIP officials allocated plots in squatter settlements and planned where houses could be built and by whom. In reference to the squatter relocation policy, following the introduction of a one-party state in 1972, Mulwanda and Mutale (1994) point to the fearless manner in which UNIP attempted to enforce town planning regulations by demolishing illegal settlements. However, the re-introduction of plural politics in 1991, has seen a shift in the relative balance of power with squatters wielding more political leverage over the ruling Movement for Multi-party Democracy (MMD). The same political strength is identified as having contributed to the reluctance of the mining authority to force squatters to move from mine owned land to Twatasha site-and-service scheme between 1974 and 1975 (Tipple, 1978).

An example of the informal sector as a factor in influencing urban morphology in Kitwe is illustrated by the manner in which Chisokone 'B' market was established adjacent to the old market and adjoining the CBD. This market as it stood in 1994 was, and probably still is, an illegal structure on land designated for commercial stands. According to Evaristo Onani, the organizing Secretary and himself a marketeer of the same Chisokone 'B' (ZNBC, 'Points of View' programme 9 September 1994), the whole venture was inspired by the Presidential directive not to harass street vendors. The vendors, empowered by this directive organized a committee which worked with the Kitwe City Council and this was followed by a demarcation of the stalls using a plan prepared by the council. The vendors' committee proceeded to allocate plots on behalf of the council. At a meeting with the vendors' committee, the Kitwe Mayor, in a bid to persuade the vendors from occupying Chisokone 'B', told the committee that the council had plans to build a market complex. To which revelation the committee suggested Chisokone 'B' as the ideal place. To date, there are no indications that the market will move. A task force is in place and this works closely with the civil police to ensure security. Although at the time of this field work (September 1994) the marketeers at Chisokone 'B' had not been licensed, the City Council had asked them to formally apply. While acknowledging the illegality of the premises, the Town Clerk argued that considering the hardships being faced by the people, the City Council could not simply move in to pull down the structures. Daily monetary collections were made by the City Council and these levies were regarded as a fine for the breach of council bye-laws. Thus, we see that while formal planning powers still vest in the City Council, political expedience and humanitarian considerations have been predominant factors in controlling urban development especially in the post-independence period.

Conclusion

This longitudinal study of the function, spatial and organizational aspects of the development of Nkana-Kitwe has contributed to understanding the extent to which colonial institutions with economic and political interests contributed to shaping the physical form and the institutional structures in existence in 1999. The mining company and the settler business community with strong economic interests and the Territorial government with both political and economic interests were the initial power bases in the development of Nkana-Kitwe. The related spatial pattern resulting from this distribution of economic and political power betrays the value judgements about the kind of society those in power wanted to create. By the use of cleverly devised instruments, a racially segregated city was developed, the sprawl of which was exacerbated by a broken terrain. As a result, Nkana-Kitwe exhibits a highly compartmentalized form with each compartment contained within some artificial or natural fences. The lack of contiguity in the city's physical form must add to the cost of services.

As regards the perpetuation of power bases, there seems to be a shift in influence, with political and executive power having a dominant role in shaping the city's development after independence. Whereas executive power, in the form of the local authority, controlled the growth of the formal sector, political power concentrated its influence in the informal sector. The demarcation of influence has not always been clear cut, occasional conflicts have occurred and political attitudes too have changed with changes in the country's political economy (Mulwanda and Mutale, 1994). The independence of the local authority from politics ended with the 1963 local government elections when councils were elected on party tickets. After independence, politicians and council executives became the dominant players in local authority, replacing the economic power of the trading community, who had contributed to the development and management of the city. This displacement of economic power by political and council executive power was not complete, but only succeeded in supplanting the small settler economic power. The more formidable mining company continued to exist not only as an influence on Kitwe, but also as a planning authority in Nkana. Although the original capitalist aim of ensuring maximum control over its workforce has disappeared with the nationalization of the mines and despite the call to integration in the Local Administration Act of 1980, Nkana Mine Township continues as a specified residential area within the city under the management of the mining company and still provides an opportunity for controlling and influencing development in Nkana-Kitwe. With the re-privatization of Nkana Division still being negotiated, any discussion about the sort of changes in the administration of the town this re-privatization will bring is merely speculative.

This chapter has also shown that after independence, the ability to influence planning decisions is no longer the preserve of formal power structures, but rather it includes both the formal and informal structures which have organized themselves around common political or economic interests. Power, therefore, should not be seen as a monopoly of formal structures, as has been illustrated by the ability of the informal sector who have no seat on the city's government but have managed to manipulate the establishment and influence the physical development of Nkana-Kitwe.

With central government holding 60.3 per cent of the shares in the mining industry before the on-going re-privatization programme (Bull and Simpson, 1993), the local government cannot use political leverage to extract economic favours from the mining company, neither is it necessary for the mining company to retain a good corporate citizen image. This and the poor performance of the mining industry since the mid-1970s has resulted in the situation in which the mining company actually welcomes the shedding of this social burden. On the other hand, local government has not been very keen to take on the task of fully taking over the management of Nkana Mine Township given the former's poor financial standing. This combination of factors might lead to the re-emergence of small private capital working in partnership with local government. The co-option of leading business people on the council's occasional committees, appeals by the city mayor to local businesses for help in the maintenance of the city, the formation of residents' associations are the beginnings of this re-emergence. Several neighbourhood associations exist in Nkana-Kitwe working with the police to maintain security. The past president of the Kitwe District Chamber of Commerce and Industry was co-opted onto the council's Investment Committee in 1995. In addition, business houses have also contributed to lighting up Kitwe's streets.

Although the multiplicity of power bases has the potential advantages for checking abuses and to offer a rich mix of alternative structures of development and organization, the realization of these advantages is dependent on the relative balance of power and the degree of co-operation between the various power bases. It also has disadvantages, as can be seen in the structure of Nkana-Kitwe, where one logical planning unit was split between two authorities, thus introducing a break in the planning statement. The actual definition of public and private roads was a source of argument between the Territorial government and the mining company, with the latter insisting on the privacy of its roads and hence their being capable of being closed to the public at any time without prior notice. The other conflict concerns infrastructure standards (a subject taken up in Chapter 7), which are generally higher in the mine township, which highlights the pitiable state of the public township's services and thus puts pressure on a cash strapped local authority.

With reference to the spatial aspect of Nkana-Kitwe, there is a relationship between the spatial structure and the distribution of power between various groups. The initial distribution of power was shared between the mining company and the colonial government. Whereas the mining company's sole objective was economic, the colonial government had the twin objectives (economic and political) of balancing the demands of the mining company and the protectorate role in relation to the local people. In pursuit of this economic objective, the mining company managed to exert its influence on the Colonial Office resulting in the establishment of closed segregated compounds which facilitated the company's desire to maintain a high level of control over its workforce. Similarly, the differentiation within functional and residential areas of Kitwe's formal development reflects the ideas and attitudes of the colonial planning profession and strong government control. Here, too, existed residential areas separated on racial lines and clearly defined functional zones. The influence wielded by small private capital was strong enough to win some limited monopoly in trade and perhaps indirectly influenced the structure of the city through its elected representatives on the management board. But, by and large, it did not have a direct influence on the spatial structure of the city. Compared to the mining company, this group of small private traders was in a weak position to the extent that it could not prevent the Kitwe Municipal Council providing additional commercial plots in the town centre contrary to the undertaking by the Territorial government not to do so.

At independence, Nkana-Kitwe's linear structure had been established. The Kafue river formed the eastern limit and the copper ore precluded further development to the west. Although the overall linear structure has been sustained by the Zambian planning authorities following independence, the detailed development of the city exhibited elements of fracture in the period 1966-75. What happened is that the development of site-and-service housing areas was in the majority of cases dictated by the siting of the squatter areas who were to be beneficiaries of these serviced areas. What is evident in this post-colonial period is not a well planned extensive development, but rather, a hotch potch of site-and-service areas and other low and medium-cost housing. This is another example of the power of the informal sector to influence planning decisions in the growth of Kitwe. Evidently, the initial linear city structure and segregated residential and functional use pattern of Nkana-Kitwe established by the colonial planning authorities has been perpetuated by successive power bases, at least until now, except that social differentiation has replaced the racial distinction between residential areas. There has been very little urban development across the Kafue River, although this was one of the proposed direction of expansion for the city. The same cannot be said of the detailed development within the city. Post-Colonial socialist policies have resulted in a measure of integration between the high-cost and low-cost housing. Site-and-service areas whose siting have been influenced by squatter settlements have contributed to the break in the logical development of the city.

That the first distribution of power and their respective spatial structures have not completely survived in their original form is evident. The original intention of the colonial planners when they said 'Come, Let Us Build Us a City' was a city whose spatial structure, functions and character served their needs. The reality today is that Nkana-Kitwe is being adapted to serve the functions of a population it was not planned for. As one moves from the CBD to the edge of the city, the city structure collapses and there is no morphological unity. The reason why the city's power bases and structure have not perpetuated fully could be attributed to the fact that these colonial minority power bases lacked legitimacy and were held in place by force. Structural changes followed mounting social pressures which had culminated in independence in 1964. Because of the tenacity of natural features and physical infrastructure, subsequent development has tended to follow the original form. But, where there has not been the compelling elements, the original structure has not guided later development. This departure from the original structure has also been promoted by the removal of controls and the rural-urban migration following the post-colonial right to movement and residence without the corresponding response by the city's planning authorities.

In theory, at least, the task of administration was supposed to be relatively easy as the mines ensured control of Nkana and the local authority that of Kitwe. The reality, however, is that strict control never led to lasting peace. Nkana-Kitwe was part of the labour disturbances which had swept the Copperbelt in 1935 and 1940 (briefly referred to in Chapter 4). Popular opinion is necessary for a legitimate and lasting administration. An administration by the minority over the many against popular opinion may be forcefully sustained and for a time exhibit a semblance of order, but it will be a purely negative policy which must reach a conflict threshold and break-up. An example of this negative policy is the disenfranchizement (alluded to earlier in this chapter) suffered by the majority African population in local government either because they were in mine townships or because they lived in non-rateable property (a definition which included all African houses). The adverse results of this exclusion are that the African was denied an early opportunity in the art of local government and was also denied the chance to influence the course of development in a city which he was eventually going to claim as his own.

In proposing the need for moral principles in urban management (see central argument in Chapter 1), it was also argued that such principles need to recognize the existence of conflicting forces representing the self-interest of groups or individuals and balance these in a way that promotes a creative tension for the benefit of all interest groups. This chapter has identified the main interest groups at various stages of Nkana-Kitwe's development and how the early tension between the mining company, Territorial government and small local business community led to the creation of the public township of Kitwe.

In summary, this chronological in-depth analysis of the evolution and development of Nkana-Kitwe reveals that urban growth and management are both affected by the political and economic interests and involve a careful balance of power between various groups, each galvanizing around one or both interests. Urban management policies which fail to identify and negotiate between contradictory interests are bound to fail. Observe that at every stage of Nkana-Kitwe's development, it is the distribution of effective political and economic power which determines the growth of the city. Where this relationship breaks down, it is because of political compromises made to maintain social stability. Planning, development and management of a city should also recognize the existence of informal structures of power, their needs and acknowledge that success will depend on balancing the needs of all power bases. To treat the

Chapter 6

Supply of Land and Property and the Changing Structure of Rates

Introduction

Urbanization puts pressure on land for commercial, industrial and residential need, and on the local authority to provide public services. The management of land as a platform for urban development and as a source of funds for public services becomes critical in urban areas. This chapter on land and property is linked to such political, social and economic issues as its supply to different groups, and the levying of rates to different groups of owners for the supply of public services. A distinction is made between 'property', which is used here to refer to developed land or the improvements on land, and 'land' which refers to undeveloped land. Where the term 'owner' is used in relation to land and property, it is intended to refer to persons or institutions whose names appear in the Land Register or valuation roll.

The first section identifies and examines the main forms of supply through which institutions and individuals access land and property. Supply may be classified as formal, when allocations are made by the government to the public, and subsequently between private individuals and/or institutions. The former mode of supply is classified as a government allocation and the latter as a market allocation. Supply is classified as informal in reference to unauthorized acquisitions of land.

The second section is a spatial and temporal analysis of land and property values. Summary information in the form of a map and index-linked average values have been used to explain variations in space and time of land and property values.

Land and property still constitute popular sources of local authority revenue for financing public services. The third section traces land acquisition charges for the provision of social and physical infrastructure, and the rating system and how rates have been distributed by the owner group. The economic relationships between institutions and/or individuals are investigated for equity in the funding of public services.

Sections four and five respectively evaluate the rating system and the 1975 Land Act's impact on the land and property market. Based on the critical appraisal of the rating system, a number of problems are identified. Section six offers a range of recommendations, while the concluding section seven draws together findings and relates these to the theme of equity, arising from the central argument that urban management decisions affect people's lives through the allocation of benefits and burdens.

This chapter draws on three main sources of data. The section on the supply of land and property is based on data abstracted by the author from the Kitwe Township Land Register. The discussion on values and rates draws upon Kitwe valuation rolls obtained from KCC and the Government Valuation Department (GVD). Archive sources have also been used to provide background information on the origins of rates in Kitwe, and some information on the contribution of rates to local authority revenue in Kitwe, especially between 1960 and 1964.

Supply of Urban Land and Property

Access to urban land and property in Zambia has been through formal acquisitions from the government or quasi-government organizations, or through subsequent private transactions. The first form is hereafter referred to as a government allocation and the second as a market allocation, both being collectively classified as formal allocations. A third and informal supply of urban land is characterized by clandestine allocations mainly by politicians, or forced occupation of private or public land, both resulting (although not necessarily) in squatter settlements.

Before 1960, leases could only derive directly from the Crown, but the Crown Grants Ordinance of 1960 empowered local authorities to sub-let land for which a head-lease had been issued by the Crown to them. In discussing government allocations, no distinction is made between Crown/State allocations and local authority allocations. Before land is released by the government for public allocation, the traditional approach (at least in theory) has been to plan, survey, service and then allocate (part of this procedure is described in Chapter 4). The main urban land use classifications in Zambia are residential, commercial, industrial and open spaces. This discussion is limited to the first three. Although urban agriculture has been researched by Sanyal (1981) and Rakodi (1988a), it is briefly discussed in relation to residential plot sizes in Chapter 8.

All dealings in land are required to be registered in the Lands and Deeds Registry. These include: grants of land, conveyancing or transferring land or any other interest in land, lease agreements for more than one year and mortgages. Three separate registers are kept (open to the public on a payment of a search fee): the Township Lands Register (for all land within townships); the Lands Register (for all land outside townships); and a Miscellaneous Register for all other documents. The information in the Lands Registers include the registration number of the document, date the document was lodged and registered, property number, names of parties to the transaction, amount paid, nature of document issued and the area of the land parcel as surveyed.

For this research, all entries in the Kitwe Township Land Register (hereafter referred to simply as 'Land Register') were transcribed stand by stand. This information included all original grants from the government to individuals or corporate bodies of surveyed township stands (classified as government allocations); subsequent transfers (classified as market allocations); considerations; year of transaction; acreage; and any charges or mortgages affecting the land.

This information excludes unreported land transactions. While complete on original grants of surveyed township stands, subsequent land dealings might not be registered for the purpose of evading the property transfer tax. Since leases less than one year do not require registration, it is possible in this way for land to pass from one corporate body to another, if the name of the corporate body is also passed on. In other cases, private arrangements are made between the vendor and buyer to have a false value declared, declare the sale as a deed of gift or to simply ignore the registration requirement. This source further excludes most transactions in unsurveyed properties such as the site-and-service areas, dealt with through the city council. The risky nature of holding valuable land and property in someone else's name limits the number of people prepared to go into private arrangements, so unregistered transfers are infrequent and have little significant effect on overall land ownership patterns. Capital values in the registry may, on subsequent transfers, be lower than the actual values at which the property changed hands, and the temptation to make false declarations will be higher if the tax rate is high.

Data from the Land Registry has been used to establish changes in land ownership, and to analyse the shifting importance over time of government and market allocations. Land Registry data has also been used to explore speculation on transfers made before and after 1975 - speculation prompting the 1975 Land Act. Below is a detailed analysis of the supply of land in Kitwe based on sample evidence from the Land Registry.

Most housing provision in Zambia has been institutional, comprising local authority and private company (including the mines) housing. Local authority housing was predominantly low-cost and largely provided for the African population. Houses in low-density areas of the public township (Parklands, Kitwe Central, Riverside) were mostly company-owned or individually-owned. The local authority did, however, build high-cost houses in Riverside, and later Riverside Extension.

Because the government was not and still is not in the property business, it is safe to assume that all government allocations relate to undeveloped land, and subsequent private transactions could relate to either developed or undeveloped land. With reference to Figure 6.1, up to 1940 a sample total of nine residential plots had been registered as direct leases from the Crown with no private allocation, thus representing 100 per cent government allocations. As a percentage of the cumulative total of both government and market allocations, government allocations declined rapidly from 100 per cent in 1940 to about 54 per cent in 1970 and remained relatively unchanged until 1985 when it began to increase. At the time of this field research in 1995, government land releases accounted for about 56 per cent of all residential land and property allocated in Kitwe since the town was established in 1935.

The proportionate share of market allocations rose rapidly from nothing in 1940 to nine per cent in 1945, stabilized between 1945 and 1950, and rose sharply between 1950 and 1955. A modest increase was recorded from 1955 to 1960 before what could be described as the last notable rise in market allocations between 1960 and 1970. By 1970, the proportionate share of market allocations at 46 per cent nearly matched the government share of 54 per cent. Between 1970 and 1985, the market share of residential land and property allocation remained static before falling slowly and rising gently again after 1990. The period 1975-85 represents a lean period when very few land and property transactions took place in Kitwe. Although the market share nearly matched that of the government after 1970, the government has continued to be the dominant supplier of medium- and high-cost residential land and probably to a lesser extent residential property.

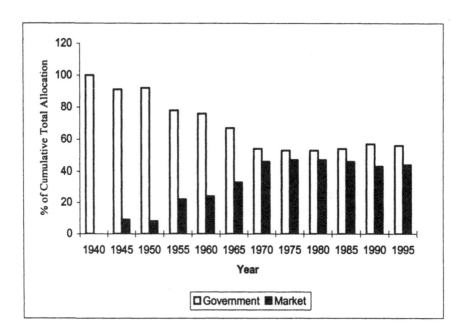

Source: Compiled by author from Land Registry data, Ministry of Lands, Lusaka, Zambia.

Figure 6.1 Relative government and market allocations of residential land and property, 1940-95

With respect to commercial land and property (Figure 6.2), the government allocations share of commercial land has steadily declined. After 1970 however, the share of all commercial land and property registered as a result of private transactions surpassed that of the government, and in 1995 the proportions stood at 62 per cent and 38 per cent respectively.

The shortage of publicly supplied commercial property is confirmed in that in the 34 years between 1958 and 1992, the sample shows that only eight commercial plots were surveyed, prompting KCC to re-zone a section of the high-cost residential area adjacent to the CBD from residential to commercial in 1990. By 1995, however, only a few of the re-zoned properties had been re-developed for commercial use. One possible reason for the slow response is the decline in mining activity which has had a knock-on effect on the level of economic activity in the city. KCC left it to intending developers to negotiate the purchase of residential dwelling units in the re-zoned area with the individual owners, leading to haphazard unplanned development. KCC reported that of 408 commercial land applications received in 1995, only six were allocated plots (KCC, 1995d).

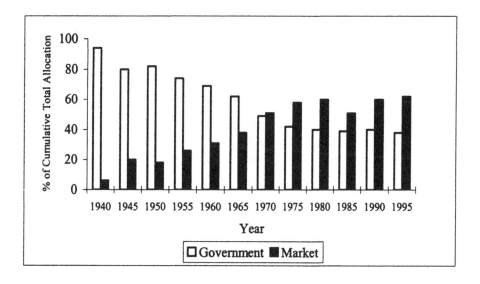

Source: Compiled by author from Land Registry data, Ministry of Lands, Lusaka, Zambia.

Figure 6.2 Relative government and market allocations of commercial land and property, 1940-95

Figures 6.1 and 6.2 compared show that the market supply of both residential and commercial property tended to match the public supply. Whereas market share in the supply of commercial property matched and exceeded that of the government after 1970 (Figure 6.2), that in residential use remained unchanged as did the associated government share (Figure 6.1). Few transactions are registered after 1975.

Until 1955, no land was registered as industrial (Figure 6.3), reflecting the government position to develop Kitwe as a commercial, not an industrial centre. In 1948, secondary industrial development was allowed for the first time in Kitwe, but it was not until 1955 that the first government allocation was registered. Apart from the disparity between the market and government allocated land and property, it is difficult to discern an established trend in the supply of industrial land and property. Government allocations constituted over 65 per cent of the cumulative total industrial land and property allocations at any one time between 1955-95, while those accessed through private transactions ranged between 15-35 per cent. Although the market has commanded a reasonable share in the supply of residential land and property, a majority share in commercial, its proportionate share of total allocations in industrial use has been poor.

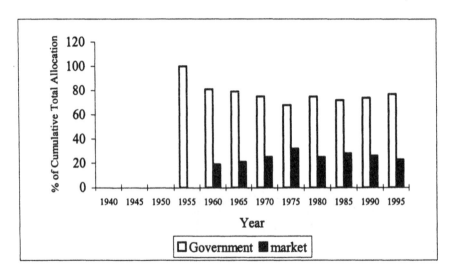

Source: Compiled by author from Land Registry data, Ministry of Lands, Lusaka, Zambia.

Figure 6.3   Relative government and market allocations of industrial land and property, 1940-95

The discussion so far has been limited to the commercial, industrial and high-cost residential sector. The supply of low-income residential land and property is now addressed. In 1989, low-cost housing in site-and-service areas, which until then had been exempted from the payment of general rates, was included in the valuation roll (a detailed description of valuation roll data is done later). To ascertain the level of transactions in this type of residential area, the 1989 and 1996 rolls were compared. When ownership changed it was assumed that this was the result of a market transaction. Although not all such changes in ownership are the result of monetary gain, because of the relatively low number of non-monetary transactions relating to high-cost housing in the Land Registry, it seemed a reasonable assumption.

Table 6.1   **Sample level of ownership changes in Kitwe's low-cost residential areas between 1989 and 1996**

Locality	Number of undeveloped plots changing ownership	Number of developed plots changing ownership	Total change (undeveloped + developed)	Sample size	Total change as % of sample
Bulangililo	0	10	10	700	1.4
Luangwa	2	12	14	795	1.8
Twatasha	2	10	12	430	2.8
Ndeke G	0	4	4	264	1.5
Ndeke H	6	12	18	511	3.5
Total	10	48	58	2,700	2.1

Source: Compiled by author from KCC valuation rolls.

Comparing ownership changes between developed and undeveloped land (Table 6.1), most between 1989 and 1996 took place when the land had been developed already. Except for Bulangililo, which was mostly built up by 1996, Luangwa, Ndeke H, Ndeke G and Twatasha still had vacant plots, with the highest being in Luangwa. Compared to ownership changes of developed land, ownership changes in undeveloped land in low-cost residential areas were few. In the seven year period (1989-96) a total of ten undeveloped plots had changed ownership from an estimated sample total of 2,700. Although some transactions may not have been recorded, changes of ownership averaged 2.1 per cent in the low-cost residential sector between 1989 and 1996, and 2.5 per cent in high-cost residential areas (Land Registry data). Notwithstanding commercialization in low-cost residential areas, people seem able to buy a house more readily than they would vacant land. Further evidence of commercialization in site-and-service areas is provided by KCC's annual report of 1995, disclosing the registration of 26 assignments, six caveats, ten mortgages and three deeds of gift (KCC, 1995d). Both in the low-cost and high-cost residential areas, the market is slowly re-asserting itself as a tool for resource allocation. Because of the illegal nature of informal land acquisitions, data is not directly available. It is only by observing the growth of informal settlements that a proxy measure of most informal acquisitions in residential use can be made (Table 6.2).

A comparison is made of council low-cost and informal housing units in Chapter 8. Suffice to say at this point that the global proportionate share in 1998 stood at 14,907 informal dwellings and 42,000 formal dwellings. Table 6.2 reveals a general increase in the informal supply of land and housing. The question this presents and which is discussed in Chapter 8 is: Why are so many people not able to acquire housing land through the formal processes?

Table 6.2   Growth of informal dwellings in Nkana-Kitwe, 1944-98

Year	1944	1974	1981	1983	1991	1998
No. of dwellings	400	9,800	13,792	15,072	10,236	14,907

Source: Compiled by author from various council reports and other publications.

Changes in Land and Property Values

The function and purpose of the Kitwe valuation rolls is now outlined. The valuation rolls are compiled with each property entry showing the owner's name and postal address, a brief description of its use, physical address of the property (street and plot number), plot area and the rateable value. Until 1975, property rates in Zambia were levied on both land and improvements based on the capital value as opposed to the rental value. The Land (Conversion of Titles) Act of 1975, apart from converting all freeholds and leases longer than 100 years to 99 year leasehold, introduced the concept that undeveloped land has no value, and from then until 1995, only unexhausted improvements upon the land attracted rates. Thus before 1975, the rateable value would be the sum of the land value and the improvement value and after 1975, only the improvement value. Although the Land Act of 1995 re-introduced the concept of value to bare land, this change came after Kitwe's current valuation roll (1996) had been prepared. In addition, s2 of the Rating Act sets out the basis of valuation as the open market value.

Valuation rolls can be prepared by the GVD, the local authority or a private valuer (appointed by the local authority and approved by the Minister of Local Government). The Rating Act of 1976 provides for a complete revaluation at least once in every five years. Supplementary rolls may record changes between the main rolls.

Until the late 1980s, the Rating Act excluded from general rates all unsurveyed properties (notably, low-cost council housing, site-and-service housing, and upgraded squatter settlements). In addition, plant used for water and sewerage treatment, places of public worship, vacant land, diplomatic housing, railway tracks and mining institution property were also exempted.

In discussing the use of rating assessment data as a source of value maps, Howes (1980) identifies three basic advantages which have also been found to hold true in this research: firstly, the ease of access to information; secondly, accuracy and conformity of values within the statutory definitions; and thirdly, periodic re-assessment. Table 6.3 is a list of valuation rolls prepared for Kitwe between 1954-96.

Table 6.3   Summary of Nkana-Kitwe valuation rolls, 1954-96

Year Prepared	Number of entries	Year prepared	Number of entries
1954 (m)	556	1969 (s)	588
1963 (m)	2,076	1970 (m)	2,850
1963 (s)	79	1978 (m)	5,075
1964 (s)	268	1981 (m)	6,982
1966 (s)	164	1989 (m)	12,479
1968 (s)	206	1996 (m)	12,976

Notes:
1   (m) and (s) represent main and supplementary valuation rolls respectively. 1978 and 1981 valuation rolls were not available for this research.
2   The number of entries is almost certainly lower than the number of properties because of single entries used to refer to a group of low- and medium-cost houses belonging to the local authority and the mining company.

Source: Compiled by author from various valuation rolls and other reports.

The following procedure selected samples for this research. From the 1996 valuation roll, every other entry was entered into the computer using a spreadsheet, forming the template for earlier valuation rolls. Properties appearing in the template but not shown in earlier rolls (because they had not been created at the time) were deleted. Using this procedure of systematic sampling, spreadsheet files were created for the following valuation rolls: 1954, 1963, 1970, 1989 and 1996. Bias may occur when the regularity of plot selection coincides with an identified pattern in the valuation roll. For example, if high-cost houses are interspaced with low-cost at regular intervals, then it is possible that only one type of house would be sampled. No such pattern was identified, suggesting that the risk of bias is minimal. Having entered all the data on the spreadsheets, a process of data checking, cleaning and calculation followed and finally several queries were designed and executed on each of the files. Summary information is shown in either map, chart or tabular form.

Figure 6.4 shows the distribution of weighted average land and property values in 1996. Although the use of average values hides wide local variations, it provides a quantitative context within which wider issues related to urban land management are discussed. The classification of residential areas into low-, medium- and high-cost is based on plot size as defined in the City of Kitwe Development Plan, Survey of Existing Conditions, p.56.

Figure 6.4 shows that the CBD with its high accessibility generally forms the core of high property values in Nkana-Kitwe. As one moves away from the CBD, property values generally decline. The low property values of Kwacha, Bulangililo, Luangwa and Twatasha are on the fringe of the city. But, there are exceptions. Although Riverside, Kwacha and Bulangililo are about the same distance from the

CBD, values in Riverside are higher than in Kwacha and Bulangililo. In a theoretical free-market situation in which land values are related to accessibility (Hurd, 1924), sites located at a similar distance and with similar accessibility would be developed to the same value (Balchin *et al.*, 1988). In the case of Kitwe, the planners shape the basic spatial characteristics of land and property values. Although Nkana East and Nkana West adjoin the CBD, property values in both areas are significantly lower than Riverside which is further away, for three possible reasons:

(i)   values in Riverside are based upon market comparisons, but Nkana East and Nkana West are institutional houses not affected by the market;
(ii)   representation from the mining company may have resulted in these lower rateable values; and
(iii) Nkana East and Nkana West have not been developed to full potential and might point to the exemption from planning ordinances enjoyed by the mining company resulting in the council not able to enforce building clause regulations which might have resulted in comparatively better quality houses.

While the planning system through its zoning regulations has influenced the spatial location of land uses and values in Nkana-Kitwe, the presence of strong economic (mining company) and political interests might have had a limiting effect on this influence, as suspected in the case of Nkana East and Nkana West. The above paragraph has examined and discussed the spatial structure of property values in Nkana-Kitwe. In spite of the fact that a value gradient is discernible, the overall value structure is influenced by a combination of the presence of planning regulations, and political and economic interests.

After the spatial aspect of land and property values, a comparison of values from different time periods is done. Table 6.4 shows weighted average current values (not adjusted for inflation) of property from different valuation rolls in different uses. The aggregate value refers to the weighted average of all sample values in that use class. Because Kitwe was a commercial centre until 1948 when the town was opened to industrial activity, industry was mainly restricted to its rival town Ndola. Kitwe's industrial property between 1963 and 1989 (except for the CBD) was valued higher than commercial property (Table 6.4), but by 1996, commercial values were higher than industrial, probably reflecting a decline in industrial activity. As world copper prices began to fall in the mid-1970s and deteriorated further in 1986, so has Nkana-Kitwe's primary industry, which in turn has affected property values in secondary industries. What effect the re-privatization of the mines and sale of institutional housing to sitting tenants will have on the land and property market is too early to tell.

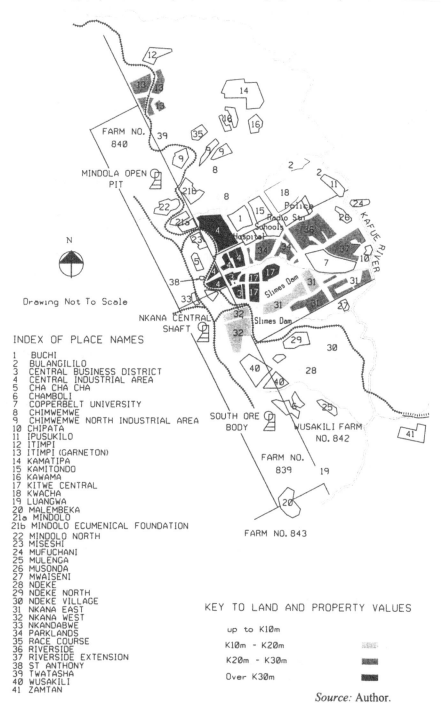

FARM NO. 840

MINDOLA OPEN PIT

N

Drawing Not To Scale

NKANA CENTRAL SHAFT

INDEX OF PLACE NAMES

1   BUCHI
2   BULANGILILO
3   CENTRAL BUSINESS DISTRICT
4   CENTRAL INDUSTRIAL AREA
5   CHA CHA CHA
6   CHAMBOLI
7   COPPERBELT UNIVERSITY
8   CHIMWEMWE
9   CHIMWEMWE NORTH INDUSTRIAL AREA
10 CHIPATA
11 IPUSUKILO
12 ITIMPI
13 ITIMPI (GARNETON)
14 KAMATIPA
15 KAMITONDO
16 KAWAMA
17 KITWE CENTRAL
18 KWACHA
19 LUANGWA
20 MALEMBEKA
21a MINDOLO
21b MINDOLO ECUMENICAL FOUNDATION
22 MINDOLO NORTH
23 MISESHI
24 MUFUCHANI
25 MULENGA
26 MUSONDA
27 MWAISENI
28 NDEKE
29 NDEKE NORTH
30 NDEKE VILLAGE
31 NKANA EAST
32 NKANA WEST
33 NKANDABWE
34 PARKLANDS
35 RACE COURSE
36 RIVERSIDE
37 RIVERSIDE EXTENSION
38 ST ANTHONY
39 TWATASHA
40 WUSAKILI
41 ZAMTAN

SOUTH ORE BODY

WUSAKILI FARM NO. 842

FARM NO. 839   19

FARM NO. 843

KEY TO LAND AND PROPERTY VALUES

up to K10m

K10m - K20m

K20m - K30m

Over K30m

Source: Author.

Figure 6.4   Average land and property values in Nkana-Kitwe in 1996

Table 6.4   Average rateable values of urban property in Nkana-Kitwe, 1954-96

Classification	1954 £	1963 £	1970 £ (K)	1989 K	1996 K
Residential					
Low-cost	--	--	--	81,150	3,151,410
Medium-cost	--	1,350	2,190 (4,370)	278,910	9,585,640
High-cost	6,540	5,520	7,230 (14,470)	594,230	23,565,030
Aggregate	6,540	5,490	7,140 (14,280)	344,130	13,442,570
Commercial					
CBD	23,870	61,300	69,880 (139,760)	5,526,120	169,334,000
Martindale	8,020	10,230	13,430 (26,860)	1,946,710	68,704,200
Aggregate	12,130	25,110	33,060 (66,110)	2,734,880	78,656,260
Industrial					
Main	--	35,000	41,430 (82,870)	2,955,780	80,829,000
Cottage	--	15,550	27,000 (54,000)	1,023,760	25,483,330
Aggregate	--	34,160	41,110 (82,210)	2,869,530	77,966,290

Source: Compiled by author from various KCC valuation rolls.

To allow for the effects of inflation, Zambia's Consumer Price Index (CPI) (Table 6.5) was used to convert the current values in Table 6.4 to real or constant values in Table 6.6. The conversion was done by dividing the current values by the corresponding CPI figures and then multiplying by 100. 1970 was chosen as the base year because it marks the beginning of substantial rises in inflation which, in the period 1954-96, peaked in 1992.

Table 6.5   National consumer price indices for Zambia, 1954-96

Year	1954	1963	1970	1989	1996
CPI	75	75	100	6,038	564,400

Note: Due to the non-availability of data for 1954, the CPI for 1960 which remained stable until 1964 is also used for 1954.

Source: Computed by author from various issues of IMF's *International Financial Statistics Yearbooks.*

Table 6.4 shows sharp rises in current values after 1970, attributable to the effects of inflation and currency depreciation (from 0.71 Kwacha to 1 US$ in 1971 to 938 in 1995) (GVD, 1996). After adjusting for the effects of inflation, Table 6.6 shows that residential and industrial property real values declined between 1954-96, while commercial values rose dramatically between 1954-63, they too were in decline after 1963. The period 1954-63 almost coincides with the period 1956-65 of increased property development in Kitwe (see Chapter 5) and the increase in commercial values between 1954-63 might be a reflection of the effect of good commodity prices world-wide during the second half of the 1950s. Due to rising prices, copper production more than doubled between 1950-60 (Turok, 1989). The increases in copper production and prices with consequent increases in incomes from the copper industry might have led to an increased demand for commercial property (office space, trade, services etc.) thus pushing up commercial property values. Additional evidence of the rising demand for commercial property is shown in the waiving of the 'closed township' agreement and the laying out, in 1953, of more commercial plots (see Chapter 5).

Table 6.6   Average constant property values for Nkana-Kitwe, 1954-96 (in 1970 prices)

Property Type (aggregate)	1954 £	1963 £	1970 £ [K]	1989 [K]	1996 [K]
Residential	8,720	7,320	7,140 [14,280]	5,700	2,380
Commercial	16,170	33,480	33,060 [66,120]	45,300	13,940
Industrial	--	45,550	41,110 [82,220]	47,530	13,810

Source: Based on author's calculations using Tables 6.4 and 6.5.

Notwithstanding that industrial property real values were 11 per cent up on their 1970 value in 1963, the period between 1963-70 shows relative stability in all property values, reflecting an equally settled economic and political period in Zambia. By 1989, real property values had fallen sharply (by an average 44 per cent of their 1970 values) with the highest fall of 60 per cent recorded for residential property. By 1996, real property values had dipped further still and were down on their 1970 values by an average of 81 per cent. One possible reason for these steep falls in real property values after 1970 is the decline of the copper industry already alluded to.

In terms of value ranking, between 1954-89, the highest values have been for industrial property followed by commercial and residential. By 1996, however, commercial property commanded the highest values followed by industrial, probably reflecting the decline in industrial activity following the fall in mining fortunes.

Undeveloped current land values for the period 1954-70 are also analysed. With the abandonment of the undeveloped land value concept in the 1975 Land Act, it is not possible to include sample data from 1989 and 1996 valuation rolls as these reflect improvement values only. Except for the steady rise in residential and commercial land values between 1954 and 1970, it is difficult to discern a pattern in industrial land values for the reasons presented below. In terms of value ranking, land zoned for commercial use is the most valuable followed by industrial and residential (Table 6.7).

Table 6.7   Average land values in Nkana-Kitwe, 1954-70

Use Zoning	1954 (£ / square metre)	1963 (£ / square metre)	1970 (£ / square metre)
Residential			
Low-cost	--	--	--
Medium-cost	--	0.28	0.30
High-cost	0.31	0.32	0.38
Aggregate	0.31	0.32	0.38
Commercial			
CBD	6.41	15.16	14.64
Martindale	0.95	1.50	1.94
Aggregate	1.95	5.67	6.58
Industrial			
Aggregate	1.14	0.46	0.94

Source: Based on author's calculations from KCC valuation rolls.

Current land values in Table 6.7 are similarly deflated by the application of the CPI to yield real or constant land values in Table 6.8.

Table 6.8   Average constant aggregate land values in Nkana-Kitwe, 1954-70 (in 1970 prices)

Use Zoning	1954 (£ / square metre)	1963 (£ / square metre)	1970 (£ / square metre)
Residential Aggregate	0.41	0.43	0.38
Commercial Aggregate	2.60	7.56	6.58
Industrial Aggregate	1.52	0.61	0.94

Source: Based on author's calculations using aggregate values in Table 6.7 and indices in Table 6.5.

Except for minor fluctuations, residential real land values remained stable between 1954-70, while commercial values varied widely, e.g. in 1954 they were 60 per cent lower than the 1970 value, rising in 1963 to be 15 per cent higher. Similar fluctuations are evident in industrial values which, in comparison to 1970, were 62 per cent higher in 1954 but 35 per cent lower in 1963. A number of possible explanations are presented for the stability in residential land values and the swings in commercial and industrial land values. The consistency in residential land values might be due to the predominance of institutional housing, which means that land in residential use was not exposed to market opinion turbulences as commercial and industrial might have. Apart from the structural effects of demand and supply affecting commercial and industrial land values, local authority policy practices also worked to affect these values, e.g. two possible factors could explain these changes. Firstly, the huge increase in commercial land values between 1954 and 1963 (Tables 6.7 and 6.8) could follow the council policy of allocating plots to the highest tenderer (City of Kitwe, 1970). Secondly, the fall in industrial land values between 1954 and 1970 could follow the policy of offering industrial land cheap to attract industrial development (City of Kitwe, 1970). While industrial land was valued at £1.14 per square metre in the 1954 roll (Table 6.7), industrial plots were reported to have been offered at a fixed rate of £0.25 per square metre between 1966-70 (City of Kitwe, 1970), dampening industrial land values. Dunkerley (1983, p.10) observes that:

> When the price of land rises much more rapidly than other prices, and the high profits cannot be justified by private improvements to the land, there is an obvious case, on the grounds of equity, for capturing part or all of these profits for the public purse...

While it cannot be established that land was cheap (as claimed by Bower *et al.*, 1997 in reference to Zambia's housing policy in the 1960s), residential land values were certainly stable between 1954-70.

Given the stability (in real terms) of residential land values (Table 6.8) and declining real residential property values (Table 6.6), the limited home-ownership can be explained by low wages and tied-housing policy. At independence, average African and European wages in Zambia were in a ratio of 1:9 (Table 5.4). The role of land cost will be developed when analysing the 1975 Land Act (see later, Impact of the 1975 Land Act) and impediments to housing in Nkana-Kitwe and Zambia (Chapter 8).

Valuing land and improvements onto a roll provides a basis for levying rates to finance public services. The next section examines the financial structure of plot premiums and rates.

Land and Property as Financial Sources for Funding Public Services
Plot Premiums: The Origins of Kitwe's General Rates System

Before rating, the financial arrangement for capital expenditure on services and general town administration was based upon plot premiums paid as a condition for the grant of a 99-year lease, and upon annual contributions (rental charge) paid into the township fund by leaseholders and controlled by the Kitwe Management Board (KMB). Under General Notice 397/35, the government and Rhokana Corporation (the mining company) were exempted from these charges:

> Persons who wish to acquire land in the new township will be required, before a lease is issued to enter into an agreement with the Management Board to provide in advance and in such amounts as may be fixed for each plot the estimated capital sum required to layout the new township and provide roads, drains, sewerage, native housing, one or more market halls for perishable foodstuffs, electricity and water mains. Annual contributions will also be required by the Management Board towards the cost of administration until such a time as the township is valued and rates imposed under the Townships Ordinance. Land acquired by the Government or by the Rhokana Corporation, Ltd. will not be liable to the charges referred to in this paragraph (paragraph 6, General Notice 397/35 : Filed at ZNA: SEC 1/1528).

In 1937, the value of premia and annual charges were respectively fixed as one-third of the value of the plot, and four per cent of the remaining two-thirds of the value of the plot (ZNA: SEC 1/1528: 20 May 1937). This arrangement worked reasonably well until 1944 when, because of financial constraints, KMB applied for a grant from the government, citing losses from sub-economic rentals on African housing and additional capital expenditure on water services. A protracted and revealing correspondence ensued within the Territorial government on the introduction of rates. Supporting KMB's application, the Chief Secretary noted that the government had made no contribution to the premium fund, and added that there was no reason why, under paragraph six of General Notice 397/35 the Management Board at Kitwe could not levy rates. If the Board had undertaken not to levy rates to premium payers, then it should carry the blame (ZNA: SEC 1/1528: Financial Secretary to Chief Secretary: 28 July 1944).

Rejecting KMB's application, the Financial Secretary responded that, while KMB attributed its financial woes to losses made on the African compound rentals and water supplies, not a penny of the premia fund had been spent on African housing, although this was an eligible expenditure under General Notice 397/35. He argued that plot premia were a transfer to KMB of funds which would normally accrue to the government as owner of the land, thus justifying the exemption enjoyed by the government (before 1960, leases could only derive directly from the Crown - see earlier, Supply of Urban Land and Property). If the government were to pay such premiums, continued the Financial Secretary, then it would 'prejudice not only its own position but that of the corporation in the inaugural scheme' (ZNA: SEC 1/1528. Financial Secretary to Chief Secretary: 17 August 1944). Since no annual contributions or rates as referred to in General Notice 397/35 had been levied, the KMB had three choices: levy the additional annual contributions; introduce rates; or, if bound not to levy rates, resign and give way to another board (ZNA: SEC 1/1528: Financial Secretary to Chief Secretary: 17 August 1944).

While the Chief Secretary, as government spokesman, conveyed the Financial Secretary's rejection of a government contribution to KMB's capital fund application, his personal views differed. In a later minute to the Financial Secretary, he says that he had been informed by Mr Coppins, the Manager of KMB, that the Board had considered it unfair to levy rates because of the high premiums paid by plot-holders and the contributions made by the same people through high profits on essential services, notably water and electricity (ZNA: SEC 1/1528: Chief Secretary to Financial Secretary: 25 August 1944). The Chief Secretary then argued for a government contribution, claiming an inequitable arrangement in which the government assumed possession of plots serviced from the premium fund to which it did not contribute. He explained that the premium fund was not made up of 'normal' premiums paid on acquisition of land (which were still being received by the government), but contributions to finance capital expenditure on roads, drains, sewerage, sanitation of the compound, market halls, electricity and water mains. It was thus only fair that the government, as beneficiary of these services to its residences and plots, also contributed to this fund (ZNA: SEC 1/1528: Chief Secretary to Financial Secretary: 25 August 1944).

Dismissing the Financial Secretary's argument that government contributions to the premium fund would compromise Rhokana's position, the Chief Secretary argued that (except for those plots and residences in existence before General Notice 397/35, which had already been serviced by the mining company itself), Rhokana was paying premia on all later acquisitions, which were not exempt from rentals or future rates. In resting his case for KMB, the Chief Secretary referred to the precedent set by the government when contributions were paid to Chingola's (another mining town on the Copperbelt) premium fund for a projected closed township (ZNA: SEC 1/1528: Chief Secretary to Financial Secretary: 25 August 1944).

The Financial Secretary counter-argued that, if plot-holders were not paying for monopoly trading rights (ZNA: SEC 1/1535: 10 August 1935), then these premia should be regarded as consideration for acquisition of land, notwithstanding that other premia had been paid for the same purpose. If freedom from competition in trade was a factor in persuading prospective leaseholders to pay the premia, the government was not bound. While not ruling out payments to the township, the Financial Secretary would not refer to such payments as premia (ZNA: SEC 1/1528: Financial Secretary to Chief Secretary: 31 August 1944).

The issue took on a new dimension when KMB was expected to service three plots in a previously unsurveyed and unserviced part of the town needed for government houses (plots 153, 154 and 155). A request for premium contributions was rejected by the Chief Secretary H.F. Wright, citing the exemption clause in General Notice 397/35 (ZNA: SEC 1/1528: 20 November 1944), but this decision was subsequently challenged by the District Commissioner (ZNA: SEC 1/1528: 1 March 1945) and the Provincial Commissioner (ZNA: SEC 1/1528: 6 March 1945). Another government officer, A.G. Williams, who had been part of the group negotiating for the establishment of the public township at Kitwe, supported KMB's application:

The application... brings up once more the question of payment of premia by government in respect of plots used for government residences in Kitwe, but the question has a special significance on this occasion because an unsurveyed area without any services is affected. This means that the Kitwe Management Board will have to produce from its own safe a sum of £850 which it will spend directly for the benefit of government. I think there will be a strong public reaction if this has to be done (ZNA: SEC 1/1528: 8 March 1945).

Williams recommended that government should, without any reference to plot premia, make an *ex gratia* grant to the KMB to meet part of the cost (ZNA: SEC 1/1528: 8 March 1945). The government still refused to pay plot premia, but the Financial Secretary conceded that the time had come to re-examine the financial structure of Kitwe, observing:

...it seems unsound to carry on local government indefinitely on the basis that capital expenditure will be defrayed from premia paid on plot holders taking up their plots, and recurrent expenditure out of excessive profits on public services, i.e. water and light (ZNA: SEC 1/1528: 27 March 1945).

The Financial Secretary urged that financial pressure be brought to bear by the government on the board to introduce rates. For its part, the government would maintain its stand not to pay plot premiums (because General Notice 397/35 exempted the government from such payments), but would pay special rates as defined in s26(K) and s26(J) of the Townships Ordinance towards the cost of services in new areas (ZNA: SEC 1/1528: 27 March 1945). While special rates were paid by the government from then onwards, private leaseholders continued to pay plot premiums instead of rates. An article on Kitwe reported:

...its residents pay no rates, ...capital expenditure, too, provides few headaches, since every resident who buys a plot - from the government on a 99 year lease and at a nominal price - pays a premium to the Board, which, in effect, pays for his section of the street and the installation of essential services (*Bulawayo Chronicle*, 26 September 1947).

With the rapid growth of the town in the 1950s, the arrangement on plot premia and special rates made it increasingly difficult for KMB to plan and service land in advance of need. Faced with an expenditure of £18,000 on servicing new business plots, school and hospital plots, KMB applied to the government for a short-term loan facility, and asked the government for special rate payments on alienation of the plot, instead of on completion of the development as was currently the case (ZNA: SEC 1/1528: November 1950).

No record was found by this author of the government's response to this application, but a representation made to the Commissioner for Local Government by the Kitwe Civic Association (complaining of a rise in plot premiums by more than 100 per cent in the two years before 1953) (Anon, 1953) suggests that KMB, probably reacting to financial pressure from the government in order to force the introduction of general rates, had increased the level of premium payments.

Although a 1951 valuation roll is mentioned (City of Kitwe, 1970), the earliest roll available to this author is dated June 1954. Notwithstanding the introduction of general rates, plot premiums continued to be levied for services such as foul sewers, open storm water drains, electricity mains, water mains and laterite roads (Municipal Council of Kitwe, dated about 1959). There is no mention of social or low-cost housing and market-halls in the above list, which were intended to be premium-funded under General Notice 397/35. Plot premiums in 1959 were charged at between £0.27 - £0.67 per square metre for residential plots and between £0.34 - £0.49 per square metre for industrial plots. While a system of competitive tendering was used to allocate commercial plots, premia for industrial and residential plots in 1970 were fixed at £0.25 and £0.41 per square metre respectively (City of Kitwe, 1970). The drop in industrial plot premiums between 1959 and 1970 reflected the policy of attracting investment alluded to earlier. Under the new Lands Act of 1995, a centrally administered Land Development Fund was created for money received for the alienation of state land for private use and annual ground rent, local authorities wishing to develop an area apply to this fund, but still seem to charge a form of premium. This appeal against the threat of repossession from a worried civil servant to the president attests to that:

> We appeal to President Chiluba to intervene and discuss the issue with the concerned minister. It is not our wish not to develop the plots, but scarcity of funds is the reason. Employers no longer give loans. City councils are slapping a minimum of K1 million service charges for plots not serviced and advise you have to pay within 30 days. How do you build in a bushy area without roads, water and sewer lines like in Kamwala South and Libala which to date are not yet serviced? (*Times of Zambia*, 25 February 1998).

The commonest form of council revenue from land and property is the rate. Unlike plot premiums or service charges (directly linked to the provision of services on plots), rate revenue benefits are largely public in nature. The next section looks at the ownership structure of rateable values, particularly the position of the mining company, and then explores questions of equity.

General Rates

Earlier sections of this chapter linked land and property values to space and time, while this section analyses shifts in the percentage share of rateable values between various groups of owners and assesses the effect on rate revenue and the financial structure of public services. Seven main owner types are defined: mining company; local government; central government; other companies (apart from the mines); individuals; parastatals; and others (charities or NGOs, clubs etc.). An attempt to sub-classify individual owners under gender and race/ethnicity was abandoned for insufficient data. Table 6.9 shows the percentage share of rateable values against corresponding owners from 1954 to 1996.

Table 6.9   Owner groups share of total rateable values (in per cent), 1954-96

Owner	1954	1963	1989	1996
Mines	16.0	27.4	27.5	26.8
Local Government*	6.3	1.5	3.4	2.3
Central Government	15.3	10.1	7.2	6.8
Other companies	29.8	36.6	22.3	20.8
Individuals	29.0	18.4	17.7	19.4
Parastatal	0.1	2.8	14.2	13.4
Others	3.5	3.2	7.7	10.5
Total	100.0	100.0	100.0	100.0

Notes: * Does not include public low-cost housing.
Source: Compiled by author from valuation rolls.

The mining company was exempted from premiums in the public township on plots which the company owned before General Notice 397/35, and from rates in the mine township, but its property developed outside the mining township and after General Notice 397/35 were subject to premiums and rates. In 1954, private companies (excluding the mines) owned the highest proportion of rateable value, followed by individuals (Table 6.9). The mining company's share of rateable values in 1954 was only 16 per cent, but with the integration of Nkana East into the Kitwe Municipality in 1961, the mining company's share rose to about 27 per cent in 1963. By 1989, rates were extended to all mining property (except those directly involved with the core activity of mining) within and outside the mine township boundary.

The extension of rates to the mining township in 1989 had the potential to improve KCC's rates position, but in practice this has not happened. Revenue from rates in 1990 was estimated at ten per cent of total council revenue (Bate, 1994) compared to 24 per cent in 1986 (before the inclusion of mine property in the valuation roll) (GRZ, 1987). In 1991, the mining company unilaterally withheld 50 per cent of the second part of the rates bill, intending to offset against this amount what the council owed the mining company in water charges (KCC, 1991a). In 1993, the General Manager of Nkana Division lodged with KCC the following points of objection to a supplementary valuation roll No. 1 of 1992 on the grounds that:

(i)   an appeal was already before the Ministry of Local Government and Housing seeking exemption from owners' rates in mine townships and plant areas;

(ii)   all services in mine townships and plant areas are provided by the mining company; and

(iii) payment of owner's rates would adversely affect the company's liquidity position and this would not be in the national interest.

Further correspondence between the mining company and KCC suggests that the former finally agreed to pay rates (as published in the amended supplementary roll of 1992) but would offset this against the latter's indebtedness with respect to charges on water supplied by the company (ZCCM, 1994).

Low-cost council housing was not subject to general rates, although a report on sub-economic housing (Jameson, 1945) lists rates as an item in the determination of house rents. Thus, it would appear that low-cost council housing was not exempted from rates, since they were expected to be included in housing rent. Such indirect rate revenue from council houses has been limited by the high incidence of rent default.

What are the implications of this for local authority finance of public services? A government report on financial relationships within the local authority and between local government and central government, published in late 1965, introduced the concept of 'one town' (*Kitwe Town Crier*, 1965b): rating was to apply throughout, and services to be funded from the same sources unlike in the past when two separate funds, one for the Africans and the other for Europeans, were operated. The result of this 'one town' concept was that the General Rate Fund (GRF) was expected to finance more services than before, when it was largely restricted to the European areas of Kitwe (*Kitwe Town Crier*, 1965b). Although the GRF did, from 1966, bear a heavier charge for community services (Table 6.10), general rating for low-cost housing was only introduced with the inclusion of site-and-service areas in the 1989 valuation roll, the latter accounted for about four per cent of the total rateable value. It appears therefore that after 1966, the low-cost housing areas of the city benefited from a source into which they paid nothing in the form of general rates until 1989, when some of these areas became rateable. Low-cost council housing rateable values account for an average of seven per cent of the total rateable value in the Kitwe valuation rolls of 1989 and 1996, and appear against KCC as the owner, which suggests that up to 1996, no individual rate bills were paid by council tenants.

Individual owner contributions to the total rateable value have not risen significantly compared to 1954, when individuals' property accounted for 29 per cent of the total rateable value (Table 6.9). It has been contended (Pasteur, 1978; Bate, 1994; Mukonde, 1994a) that the rate exemption enjoyed by low-cost housing areas limits the amount of revenue available to local authorities. Whether the call for extending rates to site-and-service areas is well founded should be weighed against the cost of collecting rates. The lower the expected increase in yield, the less economic the exercise. Mukonde (1994b, p.9) suggests the principle that 'For the council to afford to meet the cost of preparing a roll, the bill should be within 20 per cent of the increase in expected revenue for the first year'. The high amount of unrecoverable rates due to a low collection efficiency which ranges between 13-30 per cent (KCC, 1991b, 1995a, 1995b) and the cost of collection calls to question the economic viability of including site-and-service areas. Given the financial constraints of KCC, the demands of equity, and the need to instil a sense of responsibility, the solution is not to omit these areas but rather to improve on the efficiency of the rate system.

When the privatization of Nkana Division is completed and all its housing stock sold to sitting tenants, individual property owners will become the biggest rate payers in Nkana-Kitwe, which is likely to affect KCC's rate collection. When mine houses have been sold to individuals, this will greatly multiply the number of owners increasing the cost of rate recovery. A report of rates and housing arrears as at 30 June 1995 (KCC, 1995c) listed individual house-owners as owing most rate arrears, although their potential proportional contribution ranked third behind Nkana Division and other companies (Table 6.9). While the report does not list Nkana Division as being in rate arrears, this is misleading because of the disputed arrangement (referred to earlier) on water charges, which means that the council rate bill normally cancels out the water charges or is in deficit, and heavily favours Nkana Division. It is also uneconomical because the council has sometimes paid more for the water than it charged, leading it to consider constructing its own additional water plant in Ndeke (KCC, 1995a).

An Evaluation of the Rating System

Rates, payable twice a year in January and July, have traditionally been a main source of revenue for the financing of community services in Kitwe. Estimates of rate revenue and the GRF as a proportion of the total KCC revenue are shown in Table 6.10. The GRF includes revenue from government grants, personal levy, charges and licences, and general rates. Rates account for over 50 per cent of the GRF. Between 1960 and 1964, rates in Kitwe contributed an annual average of 14 per cent to the revenue account (various council reports), rising to 24 per cent in 1986 and falling to ten per cent in 1990. Lack of information precludes analysis of trends between 1964 and 1982.

Tax on land and property (also known as rates) is one of the oldest forms of local taxation, offering the following (Bate, 1994):

(i) conforms with the public finance principle of ensuring that the beneficiary of public services pay for the same;
(ii) a tax directed at the rich (property-owners) of society;
(iii) relatively easy to administer because it is based on immovable and highly visible land and property compared to other taxes based on inconspicuous sources of income;
(iv) properly designed, rates provide a substantial, stable and elastic source of revenue;

Table 6.10   Rate revenue and the GRF as a percentage of total KCC revenue, 1960-90

Year	Rates	GRF	Year	Rates	GRF
1960	15.0	28	1966	--	31
1961	14.9	26	1982	17.0	--
1962	14.1	21	1986	24.0	46
1963	13.1	21	1987	20.0	48
1964	11.5	18	1990	10.0	--

Source: Compiled by author from the following sources: 1960-64 various council reports filed at Zambia High Commission Library, London; 1982 and 1990 from Bate (1994); 1986 and 1987 from GRZ (1987).

(v)   reduces local government dependence on central government and enhances autonomy in the former; and

(vi)   unlike a poll tax which is limited by the maximum affordable by the poor, rating can be designed to take account of the different abilities to pay.

Although advanced as a strength, the targeting of a tax to property-owners, most of whom by general standards are rich, is also its weakness in that the rich are likely to be articulate in opposing taxes and other charges (*Times of Zambia*, 28 February 1994; *Zambia Daily Mail*, 3 April 1996; *Times of Zambia*, 16 October 1996; *Zambia Today*, 10 February 1997; *Zambia Today*, 26 June 1997).

Sirken (1982) offers the following five criteria for assessing any form of taxation, against which we now assess the performance of the rating system in Zambia and Nkana-Kitwe.

(i) Equity: In his submissions on amendments to the Rating Act, the Chairman of the Zambian Rating Valuation Tribunal submitted that:

> The most persistent and consistent complaint and basis for objection in every district I have been to in the last thirteen years is the lack of a co-relationship between the payment of rates to council and the provision of civic services by the councils to the rate-payers (Kapumpa, 1994, p.1).

Rates are claimed to conform to the public finance principle of 'benefactor pays' (Bate, 1994), so the rate-payers expect services from the taxing local authority, yet the Chairman of the Rating Tribunal argues that 'the rate is a tax which has to be paid irrespective of whether the council is providing the services' (Kapumpa, 1994, p.1), since the Rating Act of 1976 placed no legal obligation on the rating authority to provide services.

A High Court ruling in 1980, however did relate rates to the provision of public services (*Charliewell Kakweni* vs. *City Council of Lusaka*, cited by Kitwe City Council's Town Clerk) (KCC, 1994a):

(a) the council may, with the consent of the Minister and subject to the provisions of the Municipal Corporations Act, from time to time make and levy ordinary rate upon all assessable land or upon all assessable improvements or upon all assessable property;

(b) rate is a sum of money collected by the council for the purposes of services of a public nature as opposed to services to an individual.

Based on this ruling, it is argued that as long as reference is made to the lack of or the dilapidation in services of a public nature, the rate-payers' objections are valid.

The assumption that property-owners are rich and hence able to pay rates may be ill-conceived, because property-owners include pensioners, widows and other impoverished groups (Bate, 1994). Yet properties assessed as having the same value may be levied the same rate in Kwacha regardless of whether they are income generating (commercial, industrial) or not (residential owner-occupied).

It has already been argued that residents in low-cost housing in Kitwe, at least until 1989, benefited from community services when not paying rates, an inequitable arrangement which ceased in 1989 for site-and-service areas but still continues for the recently sold low-cost council housing.

Indeed the rating system in Zambia as a whole is inequitable because local authorities have increasingly failed to provide services. As Nkana-Kitwe's Garneton Property Owners' Association Chairman argued, 'if the money was kept in the area, the township would have developed well' (*Zambia Today*, 10 February 1997). At a national level, the Rates and Rent Payers' Association has from time to time monitored the services provided by local authorities so that payers of rent, rates and personal levy are not cheated (Northern Rhodesia, 1950; *Times of Zambia*, 30 December 1990). The system is also inequitable because of the lack of differential rating, and the 'free-riding' by some low-cost housing residents.

(ii) Economic or productive efficiency: The introduction of the zero value concept for undeveloped land in 1975 meant that rate levies could now only be applied to the value of developments on the land, called 'improvement value'. Before then, rate levies for improvements were usually lower than those for undeveloped land; in Kitwe between 1956 and 1966 respective rate levies on land and improvements averaged six pence and two pence in the pound (various council reports), encouraging land to be developed or released rather than being held unproductively. After 1975, vacant land in Zambia was not taxed and, except for the threat of repossession if one failed to develop within a given period, an owner was not under tax pressure to develop, but since the Land Act of 1995 site value rating is again a possibility. Research evidence in Kitwe's Luangwa site-and-service scheme shows that 55 per cent of demarcated plots remained undeveloped in 1996 although 95 per cent of these had been allocated between 1970 and 1989.

In Ndeke H, 50 per cent of demarcated plots remained undeveloped in 1996 of which 65 per cent had been allocated.

(iii) and (iv) Administrative feasibility and yield: Sirken (1982) observes that a tax's administrative feasibility and yield depend on the ease of assessment and collection, relative costs of collection, revenue and the number and level of exemptions offered. The use of a PC-based rating system in Nkana-Kitwe has not yet resulted in economic recovery rates; collection efficiency is between 13 and 30 per cent (KCC, 1995a, 1995b). Also the Rating Act exempts too many classes of property (Kapumpa, 1994; Mukonde, 1994a).

While real average property values in Nkana-Kitwe have fallen since 1970 (Table 6.6), the potential yield from rates could still be increased by raising the rate poundage. Greenwood and Howell (1980) observe that in most Zambian councils, in 1977, rates were still at their 1969 level and central government was unwilling to approve higher charges. With particular reference to Lusaka, Pasteur (1978) notes that rates have failed to keep pace with rises in real values. In 1965, it was reported that while the general consumer price index had risen 20 per cent since 1958, rates in Kitwe remained unchanged (*Kitwe Town Crier*, 1965b). Even when higher charges are approved by the rating tribunal, local authorities are often forced to reduce rates under pressure from rate-payers (*Times of Zambia*, 16 October 1996; *Zambia Today*, 26 June 1997). Thus, rates have not offered the desired elastic source of revenue to meet increased expenditure costs, due to strong representation from rate-payers, limited new development and the general lack of real growth in rateable values.

(v) Political acceptability: Central government may accept rating as a source for local authority revenue, but not support increases. In October 1994, five local authority valuation proposals were rejected by the Rating Valuation Tribunal (*Times of Zambia*, 4 October 1994).

Assessed against these criteria, the rating system as presently administered in Nkana-Kitwe and Zambia, despite its official popularity in local authorities, is inequitable, inefficient and uneconomic to administer. The yield, even in bigger authorities like Nkana-Kitwe, is poor and the real potential total revenue has fallen in most councils since 1970.

Impact of the 1975 Land Act

In 1975, the far-reaching Land (Conversion of Titles) Act nationalized land by the conversion of all freeholds to leaseholds; ended private real-estate business; declared that undeveloped land has no value; and introduced government controls over dealing in land and property. The primary motive of the 1975 Land Act was to curb speculation. President Kenneth Kaunda cited a case in Lusaka (referred to in

Chapter 4) when a plot of land in Lusaka's CBD was re-sold for an astronomical profit shortly after being acquired.

Research evidence is inconclusive about the rate and incidence of speculation in Nkana-Kitwe. Based on information from the Land Register, between 1939 and 1992, a sample total of 31 developed and undeveloped plots changed ownership for value shortly after being acquired, but one cannot determine conclusively the motive behind these transactions. Also one cannot estimate the number of plots being hoarded for speculative purposes. It is commonly believed that speculation increases the price of land and property (UN, 1973; Van den Berg, 1984; McAuslan, 1985). While inflation-adjusted commercial property values rose sharply between 1954-63 and have fallen since 1963, residential and industrial values have been falling since 1954 (Table 6.6), thus not supporting the rapid increases of land and property values which might point to the incidence of speculation. Speculation may also increase the amount of land and property on the market, and because of the law of demand and supply, values are likely to fall (Mulenga, 1981). Again evidence up to 1970 does not support falling land and property values due to speculation; up to 1970, we see a drastic rise in commercial values and only minor falls in residential and industrial values (Tables 6.6 and 6.8). The dramatic fall in inflation-adjusted values which might be explained by the structural effects of speculation (demand and supply) appear after 1970 suggesting that the 1975 Act failed in its intent to curb speculation. It is difficult however to be categorical on the causes of these falling prices after 1970, while increased supply due to speculation may have been a factor, it is equally true that limited mortgage finance and poor wages could not support increases in property values hence the fall in property values.

Evidence collected by the author from the Land Registry reveals a number of transactions of suspicious authenticity. For example, on 16 February 1955, Rhodesia Anglo American Limited assigned to its sister company Rhodesia Anglo Mine Services about 20 properties each valued at £80,838 at a time when average rateable values were about £6,500. It is possible that the figure of £80,838, given as a consideration for each property in the Land Register, was in fact a block payment for the 20 properties, in which case each was sold for about £4,000. If not a block payment, was this a strategy by the two sister companies to manipulate the market by inflating prices through what might have been fictitious sales? In another transaction, Shell Company Rhodesia Limited on 23 December 1963 sold two residential properties on plots 616 and 753 in Parklands to Shell Northern Rhodesia Limited for £639,609 each when average rateable values were about £5,500, probably as a means of externalizing funds a year before independence. A liquidator in 1962 assigned two residential properties on plots 959 and 971 in Parklands for £252,000 compared to the average rateable value of £5,500. Such transactions were out of the ordinary and raise the possibility of individuals and companies manipulating the property market, justifying some control over fraudulent practices. The only way the government could have facilitated its control of the few freeholds was to nationalize them in 1975, but it was probably not necessary to close down all private mortgage financial houses, because the government had enough instruments without monopolizing the mortgage finance market and closing down all private estate agents, action which limited the amount of mortgage finance available for land development.

Mulenga (1981) refers to the stifling effect of the 1975 Land Act. Research evidence from the Land Registry shows that Kitwe had very few freeholds even before nationalization: only 17 plots were held in freehold between 1961 and 1975 with no freehold registered before 1961 (except for the special case of farms 839, 840 and 842 belonging to the mining company). Most land was held under lease from the state. In spite of this, the market thrived, increasing its share of land and property allocations from six per cent to 58 per cent between 1940 and 1975 (Figures 6.1, 6.2 and 6.3), suggesting that it is possible to have a market system operating within a public-owned leasehold system. The two (market and nationalization) are not necessarily mutually exclusive. A study on land markets in Uganda similarly demonstrates that 'while a freehold tenure system may foster the evolution of land markets, it is not a necessary condition for their emergence: land markets operate under a variety of tenure regimes' (Marquardt, 1995, p.5).

Even with most of the land held on leasehold, there was mortgage finance available to support the market. In 1950, when, apart from the mine held land, there were no freeholds at all, a total of eight per cent of residential properties and three per cent of commercial properties were mortgaged (sample data from Land Registry), a level sustained until 1955, after which it declined steadily. Between 1975 and 1995, only one per cent of residential properties were mortgaged. In the commercial property market, between 1975 and 1995, only one per cent of the commercial property sample had a mortgage registered, rising to two per cent in 1990 and falling to one per cent in 1995 (sample data from Land Registry). Did the 1975 Land Act stifle the land and property market to the extent to which it converted existing freeholds to leaseholds? No. We have already argued that the market was already active and growing in Kitwe when most plots were held on leases. The provision to close all private real estate business did, however, limit the amount of credit available for urban land development. Although the market continued to operate as a means of land and property allocation after 1975, under the control of the state, finance has been limited because after 1975 only the state-owned Zambia National Building Society (ZNBS) was allowed to deal in real estate business.

Although most countries committed to a land and property market advocate for freeholds, 'the property market is about rights and not in land', admits Denman (1964) one of the avowed protagonists of a free market. Denman admits that it is the rights (leasehold or freehold) which are traded, not the land. Whilst money-lenders would rather deal with freehold rights, the market is not necessarily subdued because of leasehold rights *per se*. Money-lenders and individuals or corporate developers are willing to deal in leaseholds as long as there is tenure security, which as well as being a legal category, is just as fundamentally a matter of the state of mind of the persons concerned (Doebele, 1983). Bower *et al.* (1997, p.4), on squatter rights in Zambia, observe that 'The security of occupation was greater in practice than in law and the passing of the Housing (Statutory and Improvement Areas) Act in 1974 was expected to increase *de jure* security'. Although normally regarded as second-best to freehold, leasehold tenure, where it is accepted and properly administered, has several advantages (Bullard, 1993).

Land Registry data shows that the first public allocation was registered in 1937 and the first market allocation recorded was in 1940. Although individual property investors were prepared to risk their money, only in 1949 was the first property mortgage registered, a time lag attributable to various factors: firstly, the property market needed to build up to economic levels as far as mortgage financing was concerned; secondly, the labour riots of 1940 might have eroded investor confidence; thirdly, the lack of mortgage registrations between 1937 and 1949 could have reflected the dominance of the mining company, whose financial strength did not need mortgage finance.

This discussion is relevant when a country is anticipating far-reaching changes to its land policy. Policy makers need to be aware of the depressing effect of such change because of the time taken to implement change, and by the wait-and-see or surge attitude adopted by the people. A stable and predictable policy regime may be preferable.

Improving the Rating System in Nkana-Kitwe and Zambia

What follows are practical recommendations to improve the operation of the rating system in Nkana-Kitwe and Zambia.

(i) Broadening the tax base: This may correct the current unfairness in which some developments are not rateable despite the demands they make on public services. It is proposed that no wholesale exemptions be given, only means-tested rebates or concessions.

(ii) Linking rates to the provision of public services: Rate-payers expect services to be provided from the money paid to the local authority. This relationship would promote transparency. Local authorities would be justified in refusing or withdrawing services from the section of the community not meeting its rates obligation. Community participation and differential rating would help minimize non-payment and allow for cross-subsidies.

(iii) Rate-payer community participation: This helps ensure that the rate matches the cost of services wanted by and affordable to the community. Presently, participation is limited to the inspection of the completed roll, and appeals. Local authorities should liase with communities to agree on the type and level of services needed, explain costs involved and the operation of the rates system as one way to fund such services. Community participation enhances transparency, promotes

good governance, and, for the community, engenders a sense of belonging and ownership of services. People are generally under personal and community pressure to pay, and it is easier for the local authority to exact payment for charges on services agreed with the community.

(iv) Differential rating: The charging of a single rate in Kwacha between income and non-income generating property is inequitable, and contributes to poor collection efficiency, especially when set at a level unaffordable to most people. Differential rating allows the rate to take account of actual or potential revenue generated by different developments. It can improve collection efficiency by allowing the rate to be set at an affordable level for the respective community of rate-payers. Differential rating can cross-subsidize public services, a departure from the principle of 'benefactor pays', justified by the basic nature of some services and the possible negative effects of excluding sections of the community. Those who expect exclusive use of facilities should not think that they will escape the effects of a communicable health epidemic starting in excluded communities, for diseases know no territorial boundaries.

(v) Grant in lieu of rates: Presently, the government grant is usually less than the actual rate bill (Bate, 1994; Mukonde, 1994a). Equity demands that the grant be equal to the rate that would have been paid if the property was privately owned. Government should show its commitment to decentralization and sustainable local government by financially empowering the latter, 'putting its money where its mouth is'.

Although the above proposals have the potential to contribute to a fairer rating system in Nkana-Kitwe and Zambia, their impact on its efficiency is limited. This second set of proposals is aimed at improving the efficiency of the system.

(i) Site value rating: The rating of undeveloped land may help bring more land into the productive development sector or into the land market, and may reduce land values. More land developed and reasonably priced land on the market should increase the number of new developments and therefore the rate base and yield. A possible downside is that it might work against the majority of the people who cannot afford to build and yet are expected to pay rates on vacant land. If an individual cannot afford to pay rates, why allocate land which will never be developed and for which rates cannot be paid? The only form of rebate suggested at this stage is means-tested and time-limited, after which the full rate must be paid. For the majority poor who cannot build, rate rebates are not practical, nor is it economically efficient to subsidize vacant land. What the poor need is a house, not subsidized rates on vacant land.

(ii) Maintenance of the roll: A well-maintained roll adds to ease of administration, assessment and collection, all contributing to increased revenue and efficiency of

the system. A close liaison between the local authority's building inspectorate, the planning section, the valuation department and the Ministry of Lands will ensure that new developments and land allocations are continuously assessed and added to the roll. Computer-based systems, such as the one in Nkana-Kitwe, make continuous updating relatively easy. If the roll is well maintained, subsequent revaluations need not involve laborious field work, new values being calculated by simply changing the variables to reflect the current situation.

(iii) Affordability: To be economically efficient and effective, the preparation and maintenance of the rating system must be affordable to the local authority, and the rate levied affordable to the rate-payer. While the final rate levy should be a matter of negotiation (community participation) between the local authority and the various rate-payer communities (assumes differential rating), the local authority should establish the range (minimum and maximum rate levy) within which to negotiate. In establishing the range, regard will be had to actual or potential revenue generated by and demands on public services imposed by different developments, general price increases and the ability of the rate-payers to afford the rate. At present, the rate fixed is largely determined by the size of the local authority budget. Having calculated the maximum and minimum rate, a cost effective appraisal is done by comparing the expected income against costs. A realistic appraisal would compare the costs to expected income, using a minimum rate and assuming a worst case scenario in collection efficiency.

(iv) Frequency of payments: Presently, rates are paid half-yearly - a considerable lump sum payment for rate-payers. A comparison between the added costs involved in sending out more bills for smaller payments and the potential increased revenue might help to establish at what frequency bills should be sent out.

(v) Co-operation of all participants: Because of the interactive nature of the rating system, its efficiency is affected by a number of factors within and outside local authority control. Some of the above proposals involve the co-operation of other institutions, various local authority departments and private individuals. An informed awareness of the importance of rates among all participants helps to ensure that everyone involved in the system, both within the local authority and outside it, contributes to its success.

Conclusion

Despite government controls on land and property transactions introduced in the 1975 Land Act, the market continued to grow, accounting for over 40 per cent of all residential allocation, over 55 per cent of all commercial allocations, and over 20 per cent of all industrial use allocations between 1975 and 1995. The market has made such progress, matching public allocations in residential use and surpassing it in commercial use, because, while the government instituted curbs on the market, no corresponding attention was paid to strengthen the public's land delivery mechanism, so that people continued to gain access to land and property through the market. A contributory factor to the poor performance of the land delivery system is the limited financial resources for servicing more land.

Although the market allocation share of land and property is significant compared with government, both failed to supply land and property to satisfy all residential needs. The estimated 14,907 informal dwellings in 1998 represented about 35 per cent of all formal housing units in Nkana-Kitwe.

In analysing the spatial structure of land and property values, the classical theory of diminishing values as one moves from the centre to the periphery has been affected by other factors in addition to accessibility. In a planned economy in which land use is determined by zoning regulations, the decision to designate one area as high-cost and another as low-cost, even though both might be at the same radial distance from the CBD, creates factors (economic, social and environmental) which differentiate land and property values. Land use controls have largely directed the spatial distribution of functional uses and value structure of Nkana-Kitwe, not without conflict, and limited by market and political pressures.

A detailed chronological examination of changes in land and property values has shown that current values in all property categories rose moderately from 1954 to 1970, then steeply after 1970, because of inflation and currency depreciation. In real terms, residential and industrial property values declined between 1954-96, and while commercial values rose dramatically between 1954-63, they too were in decline after 1963.

Current land values in residential and commercial use rose steadily between 1954-70. While industrial land values declined between 1954-63, they rose slowly between 1963-70. After adjusting for inflation, residential real land values remained stable between 1954-70, while commercial and industrial values varied widely.

Kitwe public township was established under General Notice 397/35, in which reference was made to premium payments for the provision of social and physical infrastructure. The development of the premium fund was traced, outlining its main features, sifting through arguments and counter-arguments on the government and mining company positions. In addition to rates, service charges with the same function as plot premiums are still levied by local authorities in Zambia.

The introduction of rates in the early 1950s provided an extra source of income for Kitwe local authority. Using detailed information from Kitwe's valuation rolls, changes over time in the distribution and value of rateable values have been examined, as well as the relationship between the mining company and KCC on the rating of mining property, and water charges on supplies made by the company to the local authority. The rate exemptions enjoyed by low-cost housing occupiers have been explored, and the implications discussed, especially for public service finance structure.

The interactive nature of the rating system (central government approval, grants in lieu of rates, ability of owners to meet their rates obligations) means that local authorities in Zambia do not exclusively own it, so that the provision of services is sometimes dependent on the timely co-operation and input of external government units. With such a network of participants, one cannot blame the local authority when other participants fail to meet their obligations. In rejecting the inclusion of mine property in the supplementary valuation roll prepared by the council in 1992, Nkana Division, the biggest rate-payer, appealed to the devastating effect rates would have on the national economy, raising the question: to what extent should a town serve national interests at the expense of its own?

The claim of speculation, which the 1975 Land Act sought to remove, remains unproven from the Kitwe evidence. Although some land and property changed ownership rapidly and at a profit, it cannot be said with certainty that speculation was the motive. Sister company transactions show sale prices diverging widely from average rateable values at the time, raising suspicions as to the validity of these sales. While the government might be criticized for legislating to control what might have been a problem local only to Lusaka, suspect sister company transactions suggest a need for well-targeted government controls of the market.

How does this chapter relate to the overall argument and in what way does it contribute to the whole theory on urban management? A key aspect of urban management identified in Chapter 1 is its distributive outcomes in resource (in this context, land and property) allocation.

Research evidence from the valuation rolls and Land Registry shows that the formal resource allocation process has largely served the interests of mining and other companies. Although individuals have comprised an important third group of benefactors, given the costs of acquisition reflected in plot premiums (later called service charges and annual ground rents), it is argued that the poor have largely been excluded from the process of formal land allocation. This exclusion is manifest in the growth of informal dwellings, from 400 in 1944 to 15,000 in 1998 (Table 6.2). Before independence, distribution of ownership closely followed racial lines. The indigenous Africans, because of their comparatively poor wages (Table 5.4) and the colonial 'migrant African labour theory' (Chapter 4), were excluded from owning land. Although the attempt to establish ownership patterns on the basis of gender and ethnicity was unsuccessful, it established that until 1965 there were no recognizable African names in the Land Registry, suggesting an absence of African land owners before that date. After independence, such racial inequalities translated into social inequalities, with the rich dominating the formal ownership of land (as will be discussed more fully in Chapter 8). Notwithstanding the government ownership of most land before 1975, and all land after 1975, the public system's ability to supply land to all social groups has been limited and sometimes abused by party functionaries.

This chapter has also analysed public financial management decisions, particularly services and land-related revenue. Before 1966, the financial structure was designed on the principle of 'benefactor pays'. Separate accounts were kept for

Africans and Europeans, and each groups' services were largely paid for from a fund made up of income received from the benefiting community. General Notice 397/35 provided for an economic linkage, financing African housing from the premium fund, but this seems not to have happened. After independence, and in line with the country's socialist policy of humanism, the 'one town' concept was introduced in 1966, so as to promote the creation of an integrated society. The resulting inequity of this concept has been explored and is an example of urban management political decisions impinging on the distribution of burdens and benefits.

The city thus contains competing and sometimes conflicting interests. The sometimes acrimonious relationships between the local authority, government and the mining company provides evidence for the postulated structural conflict model of urban management, with conflicting forces essentially self-interested. This is demonstrated by, for example, government's unflinching position in relation to plot premiums. Government's preparedness to benefit from services provided by the premium fund, to which it did not contribute, shows that, instead of being a neutral arbiter, the government, like any other force, has its own interests to serve. While the government might be an interested party in the question of plot premiums, within it different opinions, including an advocacy role for the local authority in Kitwe, existed, challenging the view of all social relationships as exploitative. Given that self-interest (whether the self is a corporate body or an individual) generally determines how interest groups behave, what can be done to persuade them to act in ways that benefit not only themselves but others as well? A proposed urban management model aired in Chapter 2 is one based on balancing existing tensions between groups, so as to create conditions for decisions and actions of mutual benefit.

Chapter 7

Service Provision in Nkana-Kitwe – The Case of Water

Introduction

The original intention of Kitwe's colonial city fathers when they said 'Come, let us build us a city' was that its spatial structure, functions and character would serve their needs (Chapter 5). In the ensuing city-building process, municipal engineering provided clean piped water, water-borne sewerage, roads and other services. Although primarily envisaged as a European town, the city also accommodated the African who enjoyed an enviable level of water supply compared to other traditional African towns. But, the colonial power base was of a minority nature and following independence, the African in Nkana-Kitwe inherited the city, with a good network of municipal infrastructure and services designed for a high-income European population, the standard of services varying between European and African sections. Because it is a basic public service, water has been selected as a specific case for studying issues of equity in distribution and pricing, balancing the inherent tensions between the public service and private good, and the need for economic sustainability.

This chapter is divided into two parts. The first part establishes the history of water supply for the mining company, and its relationship with KCC, and examines the physical aspects of water supply, including the distribution networks, types of water connections, consumption levels, and the Kitwe Development Plan and Water Supply Project attempt to forecast future demand. The second part analyses the equity of Zambia's Water Supply and Sanitation Sector by relating water charges to consumption and income, and explores an alternative pricing policy. Kitwe's tariff is compared with that of another city in Zambia and other pricing policies to determine their impact on the city. While the first part explores impact of urban management on the physical distribution of water, the second deals with policy decisions on pricing of water.

The methodology applied in this chapter fuses both institutional and interpretive approaches, using empirical data; the only comprehensive data on water consumption by sub-groups available to this research is the City of Kitwe Development Plan; while KCC responded to a questionnaire handed in by the author, the mining company failed to respond to a similar request for information. With such data limitations, the author estimated from available information.

A History of Water Supply in Nkana-Kitwe

A mine township was established at Nkana by 1929, and the mining company needed water for its core activity, which could thus be extended to the mining township for domestic use. The historical nature of water provision in Nkana therefore is that of a social overhead capital, not as a profit-making service in itself. Up until 1999, mine township residents were paying a fixed subsidized charge deducted at source for their water consumption. For this, and other reasons mentioned in Chapter 5, the mining company was prepared to sell water to KMB at eight pence per 1,000 gallons, and KMB re-sold the same at two shillings (ZNA: SEC 1/1528). Since under the Urban African Housing Ordinance of 1948 (Chapter 4) services to African compounds were supposed to be provided on a non-profit basis, we might assume that this substantial profit accruing to KMB was made on water supplies to the European public township and could be used to subsidize supplies to the African compound. In analysing current financial aspects of water supply, we will explore how this possible economic linkage between the historically European residential areas and African compounds might have continued.

Until 1968, when KCC built its own water purification plant, the city's main supply of water was a plant on the Kafue River, owned and operated by the mining company, then called Rhokana Corporation and later Nkana Division. As a result of increased population, the then Kitwe Municipal Council observed that the level of water supply would soon be inadequate, but in 1964 Rhokana claimed no plans to increase capacity at their plant. Subsequently, the council embarked on a water augmentation scheme about five kilometres upstream of the Rhokana water works on the Kafue River. A water plant with a maximum capacity of about 90,000 cubic metres per day (cmd) was designed for phased construction. The first phase, with a design capacity of 27,276 cmd, was completed in 1968, and together with the guaranteed supply of 22,730 cmd from Rhokana, was expected to suffice until 1972 (City of Kitwe, dated about 1967). The expansion of the KCC plant in 1973 increased the capacity from 27,276 to 81,828 cmd, which improvement was expected to provide for the expansion of the city to 1980 (KCC, 1973a). Meanwhile Rhokana expanded its plant capacity to 63,645 cmd (City of Kitwe Development Plan, Survey of Existing Conditions).

The city is now supplied from three plants, two of which are on the Kafue River, and the third is fed on the Mwambashi (a tributary of the Kafue). Of Nkana Division's output of 63,645 cmd, the mine plant supplies to KCC 13,640 cmd, the balance being consumed by Nkana Division in its operations and township. KCC's own plants have a combined capacity of about 83,000 cmd (City of Kitwe Street Plan; KCC, 1994b).

Although KCC now has its own water works, it still buys water from the mining company to supplement its own supplies, a costly arrangement given the council's poor cost-recovery rates, which has led to constant wrangling and the council considering constructing an additional water plant (KCC, 1995a).

A dual structure of water supply thus exists in Nkana-Kitwe. Before examining the physical structure and service levels, a brief note is made about effects of re-privatization of the mines. The reluctance of new private owners to assume the social responsibility of providing education, health, housing, recreation and other social services is likely to put pressure on local authorities (*The Post*, 6 April 1998). Private enterprise may contribute to public service provision if this is seen to be in their interest, and KCC and other local authorities on the Copperbelt could gain as private mining companies invest in the community. Thus, with a proper balance of power and incentives to the private mining companies, local authorities on the Copperbelt can benefit from private capital. Negative impacts are identified on the recovery of property rates discussed in Chapter 6.

Physical Supply of Water in Nkana-Kitwe

Table 7.1 shows the infrastructure of water treatment and distribution. Chimwemwe distribution point comprises two reservoirs built in 1969 and 1970 to service existing Chimwemwe township and projected industrial areas north of Chimwemwe township (Figure 7.1). Industrial growth north of Chimwemwe has not progressed at the rate envisaged when the area was being planned, and nearby low-cost housing areas have been supplied with water, apparently as unintended beneficiaries, and some Chimwemwe water is transferred to Kwacha Road reservoirs near Buchi (Figure 7.1).

Most of the council water distribution network is now more than 30 years old, and water losses are estimated at 30 per cent of production, compared with a national average of 27.5 per cent in England (Essex and Suffolk Water, 1997). In 1995 alone, 494 leakages were attended to by the council, 1,018 low water pressure cases were reported, and 19 valves and six fire hydrants replaced, all indicators of a dilapidated distribution network. Problems also exist in the mechanical installations and civil works at reservoirs and water treatment plants (KCC, 1995d).

Figures 7.1 and 7.2 show that water distribution in Nkana-Kitwe is fairly comprehensive. By 1974, Nkana's residents were all on a mains supply of one form or another, compared to 58 per cent of Kitwe's public township, and the water mains service now covers about 91 per cent of the city's population. Without excusing colonialism, the African in Kitwe benefited from some of its municipal technology extended to his area, but at a reduced standard. Figure 7.2 shows the clear disparity in the service standards between the historically African (low-cost) and European (high-cost) areas, and Table 7.2 summarizes service levels.

Table 7.1   Distribution of water in Nkana-Kitwe in 1994

Reservoir and capacity (cubic metres)	Water works	Areas served
Chimwemwe: 1x 20,400 1x 1,136	Council	CBD, industrial area, high-cost residential area, Buchi, Kamitondo, Kwacha, Bulangililo, Chimwemwe
Kwacha Road: 1x 22,730 2x 11,365	Council and Nkana Division	High-cost residential area, Kwacha, Bulangililo
Ndeke: (2x 3,410)	Nkana Division	Ndeke, Luangwa
Garneton: 1x 90 1x 226	Council	Garneton (Itimpi)
Wusakili: (1x 4,546)	Nkana Division	Chamboli, Wusakili
Nkana Division club tanks: (1x 1,591)	Nkana Division	Nkana West
Mindolo Reservoir: (1x 4,546)	Nkana Division	Mindolo
17th Avenue: (1x 4,546)	Nkana Division	Nkana East

Sources: City of Kitwe Development Plan (1975-2000), Survey of Existing Conditions, pp.198-8; KCC, 1994b, pp.8-9.

Table 7.2   Water supply type by households in Nkana-Kitwe in 1990

Type of supply	Number of households with supply	As percentage of all households with supply
House connection	14,431	29
Private standpipe	17,857	36
Communal standpipe/borehole	17,839	35
Total	50,127	100

Source: Derived from Kitwe Water Supply project, 1991, Table 3.3-1.

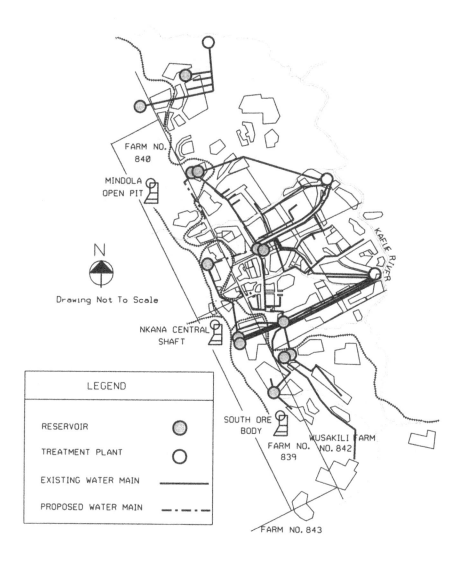

Source: Kitwe Water Supply Project, 1991.

Figure 7.1 Water supply system in Nkana-Kitwe, 1990

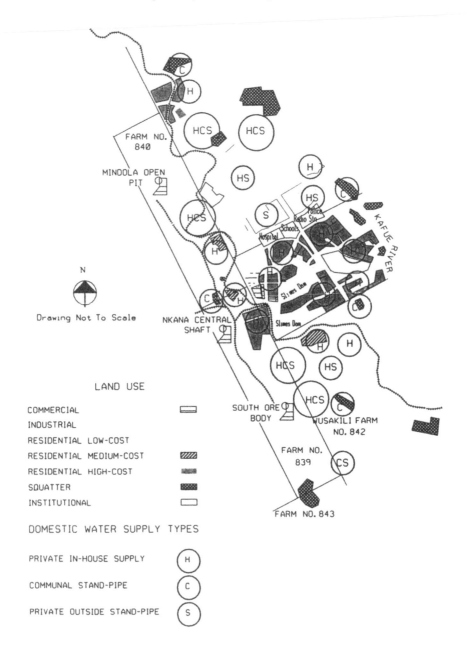

Source: Kitwe Water Supply Project, 1991.

Figure 7.2 Types of water supply connections in Nkana-Kitwe, 1990

Table 7.2 shows 65 per cent (29 + 36) of households with private supplies in 1990. Thirty-five per cent relied on communal standpipes and boreholes, most reported to be in disrepair in 1994, affecting an estimated 1,950 housing units in peripheral areas of the city (KCC, 1994b). In 1974, KCC extended mains communal water supplies to some big informal settlements (Kasongo and Tipple, 1990) and in the late 1970s sank boreholes in other outlying areas, but most had been vandalized by 1994, so that residents in areas such as ZamTan, Malembeka and Mufuchani (Figure 7.1) had to rely on river supplies (KCC, 1994b). As for those with piped water supplies, for example, Luangwa, Mulenga, Mwaiseni, Ipusukilo, Race Course and Itimpi (Figure 5.10), constant vandalism of installations is reported (KCC, 1995d) and has led the council to launch a civic awareness campaign to prevent public abuse of infrastructure. The cartoons in Figure 5.4 are examples of such a campaign. Partly as a result of such vandalism, and also due to inadequate supplies, a visit to Luangwa site-and-service scheme in April 1997 revealed that although standpipes are in place, no water runs through the system (Figure 7.3).

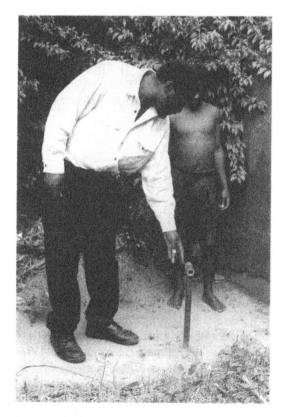

Source: Author.

Figure 7.3 Disused standpipe in Luangwa, Kitwe

Water Consumption Patterns in Nkana-Kitwe

Nkana-Kitwe's water consumption grew five-fold between 1955 and 1967 (Figure 7.4), an increase matching the increased numbers of surveyed stands (Chapter 5), and attesting to the rapid development of the town. Another five-fold increase in water consumption was then recorded, from 25,000 cmd in 1967 to 117,000 cmd in 1974.

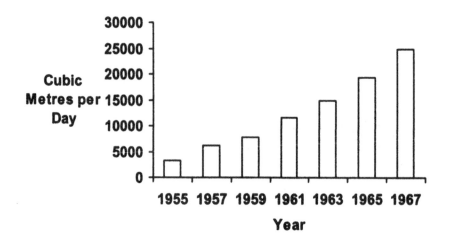

Source: City of Kitwe, dated c.1967.

Figure 7.4 Daily water consumption in Kitwe, 1955–67

Seven water tables are contained in the Appendix: Table 1 shows water consumption in Nkana-Kitwe in 1974, classified under industrial, city centre and residential (further subdivided into socio-economic groups), and consumption levels related to consumers or households. Consumer refers to the client group supplied with water. Household is defined as a group of people living and eating together; where more than one household in a house exists, the other households are referred to as sub-households.

The information in Appendix Table 1 is divided into two, relating to Kitwe public and Nkana mine township. The highest *per capita* consumption in 1974 relates to the outer suburb of Itimpi, closely followed by the general industrial area (except mining); the lowest consumption levels are in Kitwe's low-cost areas. Except for Bulangililo and Ndeke in Kitwe and most of Nkana's medium-cost housing, all of which have individual in-house connections, the rest of Kitwe's and Nkana's low-cost housing is served by private or communal standpipes. While Nkana Division supplied all of its households with piped water, about 58 per cent of Kitwe's public township population in 1974 were supplied with water. The combined population of 225,357 represents a coverage of 67 per cent.

After 1972-75, when Kitwe undertook a major planning exercise which included water supply (City of Kitwe Development Plan), the next major attempt at water planning was the 1982 feasibility study, on which no information was available for this research. The next study, the 1991 Kitwe Water Supply Project, was commissioned to augment existing plants for water demand to the year 2010, and data from this source is used. Estimates in Appendix Table 2 assume no change in the pattern of consumption since 1974. While the client population had grown, estimated consumption levels in 1990 assume 1974 consumption patterns. The increase in consumer population from 34,000 in 1974 (Appendix Table 1) to 47,000 in 1990 (Appendix Table 2) represents an increase in coverage from 67 per cent to 91 per cent for the entire population of Nkana-Kitwe. On the basis of the observed 1974 consumption rates (Appendix Table 1), the estimated demand for the public township in 1990 is 106,765 cmd. On the other hand, if Nkana's 11,076 households consumed 44,552 cmd in 1974, the proportionate increase in water demand as a result of the growth in households to 11,296 is an additional 885 cmd, raising the estimated demand in Nkana to 45,437 cmd.

Given design limitations, production and distribution losses, this estimated demand could not be, and has not been, met. These consumption estimations in Appendix Table 2 should not be taken as showing the level of consumption in 1990, but rather what might have been the total demand for water assuming water use remained at the 1974 level. Network losses, estimated as 30 per cent of production (City of Kitwe Development Plan, Final Report, p.198), reduce the combined capacity of 146,645 cmd (83,000 from KCC plants and 63,645 from Nkana Division plant) to 102,652 cmd. Compared with the estimated demand of 152,202 (Appendix Table 2), this suggests a reduction in consumption by 33 per cent.

Appendix Table 3 illustrates the drop in consumption levels in 1990 over the 1974 levels as population growth, limited investment and network losses combined to reduce the supply of water to an increasing population. It assumes that KCC can buy water from Nkana Division with no restriction, and a uniform change in consumption for all consumers. Compared to 1974, use levels drop by about 33 per cent by 1990, and of a population of 338,207, an estimated 307,894 are connected to the mains water supply leaving 30,313 to rely on boreholes and river supplies. The improved distribution level, from 34,000 consumers in 1974 (Appendix Table 1) to 47,000 in 1990 (Appendix Table 3), has been at the expense of a reduced consumption by established consumers as new consumers are added to a limited supply. Consumption levels for the public township drop from the observed 478 litres *per capita* per day (lpcd) in 1974 (from Appendix Table 1: 72,104 cmd x 1,000 = litres per day, divided by a population of 150,688 = lpcd) to 318 lpcd in 1990 (Appendix Table 3). A report on water and sewerage services in Kitwe (KCC, 1994b) observed that, due to mechanical problems, water production at council plants was only 55,000 cmd, compared to a design capacity of 83,000 cmd, before network losses; the reduction in consumption levels could thus be significantly more than the above mentioned 33 per cent (which assumes a KCC production of 83,000 cmd), as high as 45 per cent.

Owing to the lack of data on normal consumption rates in Zambia, recourse is made to recent research in the Republic of South Africa, where Behrens and Watson (1996) suggest a range of 25 lpcd to 450 lpcd with 250 lpcd as normal consumption for house connections. Compared with these figures, Kitwe public township's consumption in 1990 was high at 318 lpcd, and higher still Nkana's at 378 lpcd (Appendix Table 3). Table 7.3 shows a total normal domestic demand of 33,000 cmd, as opposed to the estimated 66,000 cmd for 1990 based on 1974 consumption levels. It is further shown that the council plant capacity of 83,000 cmd is adequate to supply an additional one and a half times the number of existing domestic consumers (from [83,000-33,000]/33,000).

Table 7.3   Comparison between estimated and normal domestic consumption

Service Type	No. of households[a]	EDC (litres/HH)[b]	ETC (cmd)[c]	NDC (litres/HH)[d]	NTDC (cmd)[e]
Communal standpipe	9,235	1,300	12,005	320	2,955
Private standpipe	15,328	1,300	19,926	768	11,772
House connection:					
(i) Itimpi	500	10,900	5,450		
(ii) High-cost	6,253	3,600	22,511	1,600	17,973
(iii) Medium-cost	4,480	1,300	5,824		
Total	35,796		65,719		32,700

Notes:
a.   extract from Table 7.5
b.   EDC = Estimated daily consumption. Extract from Table 7.5 and converted into daily consumption in litres per household (HH)
c.   ETC = Estimated total consumption. Product of columns 2 and 3
d.   NDC = Normal daily consumption. Calculated from consumption levels of 50, 120 and 250 lpcd for communal, private and house connection respectively (Behrens and Watson, 1996) and an average Kitwe household of 6.4 persons
e.   NTDC = Normal total daily consumption. Product of columns 2 and 5

Sources: Author's calculations based on Table 7.5 and Behrens and Watson's (1996) normal consumption estimates.

Water Proposals in the City of Kitwe Development Plan 1975-2000

On 13 September 1972, the Zambian Ministry of Local Government and Housing commissioned the National Housing Authority (NHA) to prepare a development plan for the city of Kitwe up to the year 2000 (the city of Kitwe at the time excluded Nkana Mine Township but included the outlying townships of Chambeshi, Chibuluma and Kalulushi which, in 1977, were excised to form Kalulushi Township). The plan identified water provision for new urban growth and improving the existing service as a major issue. Determinations of the future requirements for water supply were based on an earlier study (Gibb and Partners, 1971) which assumed 'that the demand for water will steadily increase at the rate of 2.5 per cent *per capita* per annum from 337 litres per head per day in 1970 to 707 litres per head per day by the year 2000' (City of Kitwe Development Plan, Final Report, p.198). Comparing Kitwe's 1973 population levels (Table 5.2) with the initial projection population of 1970 in Table 7.4, the plan proposed a 100 per cent supply level, i.e. the entire population connected to water mains.

Table 7.4   Development plan projections for water demand in Kitwe public township, 1970-2000

Year	Population	Projected demand for water[a]	Total demand for water[b]
1970	243,000	337	81,891
1974	*260,800*	*372*	*97,018*
1975	332,000	381	126,492
1980	440,000	431	189,640
1985	555,000	488	270,840
1990	*700,667 (actual 338,207)*	*552*	*386,768*
1995	*846,334*	*625*	*528,959*
2000	992,000	707	701,344

Note: Rows in italics are author's own interpolation to provide a basis of comparison with observed (1974) and estimated (1990) values based on observed 1974 consumption rates.

a.   in litres *per capita* day (lpcd)
b.   in thousands of litres per day

Source: City of Kitwe Development Plan, Final Report, p.198.

Taking an initial rate of 337 lpcd in 1970, an increase of 2.5 *per capita* per annum, and a 100 per cent supply level, the projected demand in Kitwe for 1974 was 372 lpcd for an estimated population of 260,800; actual recorded consumption for a population of 150,688 (representing 58 per cent of Kitwe's population) was 478 lpcd (derived from figures in Appendix Table 1). All projections in the plan seem to be based on a 100 per cent supply level, and increased demand from 337 lpcd in 1970 to 707 lpcd in the year 2000, taking no account of deficits in supply. If the 1974 water production was extended to the entire city, established consumers would have had to suffer a drop from an average of 478 lpcd to 372 lpcd. The projected plan population for 1990 was about 700,667, against an actual population of 338,207. This projected population figure even including Chambeshi, Kalulushi and Chibuluma with a combined 1990 population of 36,000 (Kitwe Water Supply Project, Vol. III, 1991) was a gross over-estimate. The plan's projected demand level for 1990 at 100 per cent supply is 552 lpcd, compared to the estimated 318 lpcd for a population of 226,723 (88 per cent of Kitwe's population).

The plan's projections can be criticized on a number of points. Firstly, the plan fails to acknowledge that the 1974 supply position was one of deficit requiring an initial heavy capital outlay before further projections could be done for a 100 per cent supply. Secondly, the plan over-estimated future population growth. Thirdly, the planned increase in demand, and its implications for production and distribution from 337 lpcd in 1970 to 707 lpcd by 2000, is unrealistically high. Although in 1974 consumption averaged 500 lpcd with a low of 266 lpcd, compared to the Behrens and Watson (1996) estimated range of 25-450 lpcd, even this 1974 consumption rate was high. Unfortunately, the plan fails to take account of the financial implications of publicly funded services when economic growth falls behind population growth, a scenario which was already evident in Kitwe in the early 1970s. The combined effect of these four points is that the plan's projections were vast over-estimates, creating a problem for which the city had limited resources to address.

The Kitwe Water Supply Project (1991) estimated the number of operational service connections in Kitwe at 6,291, which at a household occupancy rate of 6.4 represents only 40,300 people on council water mains. KCC's report (KCC, 1994b) estimates the number of service connections as 24,000. If only 6,000 of these were operational in 1991, then 18,000 were out of order!

These 1971 and 1991 reports were the work of foreign consultants, who either followed disproportionately high standards based on their foreign experience, or were driven by mercenary motives to over-estimate the problem. Standards should only be adopted where they are affordable, highlighting the importance of local knowledge to formulate practical solutions. If the population projection and consumption level estimates were realistic and the economy sound, then there would have been economic value (economies of scale in project design) in over-designing (not over-projection) to limit the costs of expanding capacity. The plan's proposals are not an over-design intended to provide capacity for future demand, but rather, it was proposed to meet the demand as it was projected to rise. Admittedly, if these proposals had been implemented, water supply should not have been an issue today but, because of financial limitations, these proposals have not been effected. A realistic approach would have been a phased increase in supply to reach a normal consumption level of the order of 250 lpcd (Behrens and

Watson, 1996) or Kitwe's own 1974 average consumption level of 500 lpcd. At a consumption rate of 250 lpcd and a total 1990 Kitwe-only population (excluding Nkana) of 257,000, the total demand is about 64,000 cmd compared to KCC capacity of 83,000 cmd. Assuming production and distribution losses (30 per cent) reduce the supply to 58,000 cmd, KCC only had to buy 6,000 cmd from Nkana Division to ensure 100 per cent supply at a consumption rate of 250 lpcd. Estimated actual consumption in 1990 was 318 lpcd for 88 per cent of Kitwe's public township population (derived from figures in Appendix Table 3). If 12 per cent of the township's population had no access to a council supply of piped water, their exclusion is largely due to the extreme high rates of consumption for certain sections of the town, and technical inefficiencies in the system. To encourage consumers to limit consumption, a system of rewards and sanctions could be structured in the tariff.

The opening section of this chapter found that water provision in Nkana-Kitwe was historically a social overhead capital, a necessary expense for the development of the mines, and consumption grew rapidly to warrant a number of studies with a view to increase supply. Although still regarded as a public service, public authorities are faced with the difficulty of revising water tariffs devised when the supply of water to African compounds was on a non-profit basis. The dilemma of urban management is to ensure availability of water to all social groups at an affordable cost which also ensures the sustainability of the service. The aim of Zambia's Water Supply and Sanitation Sector is stated to be to:

> Improve the quality of life and productivity of all people by ensuring an equitable provision of an adequate quantity and quality of water to all competing user groups and sanitation services to all, at acceptable cost, on a sustainable basis (Water Sector Development Group, 1993).

The second part of this chapter analyses the current cost recovery approach in Kitwe, exploring issues of equity and affordability by comparing income and water charges. An inter-city pricing comparison is done between Kitwe and Lusaka, and alternative pricing policies are examined.

Cost Recovery in Water Supply

Under the Urban African Housing Ordinance of 1948 (see Chapter 4), water and sanitary services were provided on a non-profit basis (Tipple, 1978), with lasting effects on cost recovery. As long as profits from beer sales (Chapters 4 and 8) in African townships supplemented rental income to finance such services, water to African townships could be supplied at less than cost. With the housing revenue account now in perpetual deficit (Chapter 8), limited cross-subsidy potential (the 'one town concept', in Chapter 6) and the failure of council commercial undertakings after independence, subsidized consumption is available at great cost to the council and the comparatively poor Zambians who live in the once European residential areas and are expected to cross-subsidize low-cost housing's water consumption. The non-profit approach to water supply is still reflected in KCC's objective of balancing expenditure on water with revenue in the formulation of its tariff (KCC, 1994b).

Water charges are levied in different ways. Residents of Nkana mine township pay a fixed proportion of their salary deducted at source by the mining company; from information available to the author, these charges are a fraction of actual cost. It has already been established that the supply of water by the mining company to its township was historically regarded as social overhead capital, on which the company ordinarily did not expect to make a profit. The council levies a fixed rate on private and communal standpipes, the former paying a higher rate. The last group of consumers levied by the council are those with metered house connections.

Figure 7.2 shows the general spatial distribution of various service levels in Kitwe's public township. For the purpose of this discussion, the 1994 tariff (Appendix) is used with occasional reference to the 1997 tariff (Appendix) which was published after this chapter was written. Table 7.5 shows the estimated consumption and related charges for the three types of consumer groups: communal standpipe, private standpipe, and metered consumption for both domestic and industrial use. The relationship between consumption and unit charges is presented in Figure 7.5. Consumption levels are estimated average monthly volume of water used by one consumer, which in residential areas represents a household of 6.4 persons. The unit charge is the levy in Kwacha (K) the council charges for the use of one cubic metre of water.

Setting aside the two consumer categories of communal and private standpipe (who pay fixed monthly rates unrelated to volume), metered consumers (Table 7.5 column 4) in Itimpi high-cost residential suburb have the highest consumption levels, estimated at 327 cubic metres per month but pay the least unit charge calculated as K258 per cubic metre. The 1994 tariff structure rewards high domestic consumption. Medium-cost housing customers, who are among the low consumers, pay the highest for each unit consumed (Table 7.5 and Figure 7.5). One possible explanation for this price discrimination is the application of a common economic practice of charging more to those customers with a low elasticity of demand, and less to those with a high elasticity. In this case, while consumers in medium-cost housing cannot reduce their consumption much lower, those in high-cost areas could in theory reduce their consumption when faced with increases in water charges (for example, not watering the lawn, less car washing or withdraw their custom by sinking a borehole). The council, to maximize its revenue, has structured a form of price discrimination based on consumption. The weighting of the tariff against low consumers is of limited moral value, whereas those singled out for high unit charges are little able to afford them. The reality, however, is different, in that most consumers in low-cost housing pay a fixed subsidized charge, so that the above situation would only arise if all domestic consumption was charged against the 1994 tariff for metered consumption.

Table 7.5 Estimates of water consumption charges based on 1990 estimated consumption and 1994 tariff

	Average unit charge (K/m^3)	No. of consumers	Average monthly consumption per consumer (m^3)	Estimated average monthly charge (K) per consumer
Low-cost communal	20	9,235	39	800
Low-cost standpipe	51	15,328	39	2,000
Itimpi high-cost HC	258	500	327	84,500
High-cost HC	262	6,253	108	28,250
Industrial	283	600	321	90,800
Medium-cost HC	320	4,480	39	12,500

Note: HC = house connection.
Source: Author's calculations.

The City Centre and Martindale consumption (Appendix Table 3) includes both high-cost residential and commercial customers, reflected in the relatively high estimated consumption rate of 4.3 cmd, compared to the average 3.6 cmd for high-cost residential customers only. Because the tariff groups commercial with industrial consumers, commercial customers receive a generous initial tariff allocation of 80 units for K18,000 (1994 tariff in Appendix), since they on average consume about 130 units per month, compared to 320 units for industrial use. The two could be separated, so that commercial customers do not benefit unduly from an initial allocation which takes account of high industrial water consumption.

Estimating that it needed K3.8bn to treat and supply water in Kitwe in 1995 (see water and sewerage questionnaire in Mutale, 1999) and assuming a full production capacity of 83,000 cmd and 30 per cent network losses, the council should have been able to raise an estimated K4.8bn from its water charges to consumers (Appendix Table 4) in 1995.

However, production is reported to have been only about 55,000 cmd, of which only 49 per cent was billed, representing a revenue of K216m per month (KCC, 1994b) or K2.6bn per year against a potential revenue of K4.8bn (Appendix Table 4, column 6). The actual amount collected was significantly lower than the billed amount due to disastrous collection efficiency, estimated as 4.2 and 3.6 per cent in March and May 1995 (KCC, 1995c). Compared with the estimated expenditure of K3.8bn (Water and Sewerage Questionnaire) there was a deficit of K1.2bn. Poor collection efficiency, because of high rates of default, political interference and KCC management weaknesses, limit the amount of water revenue collected (*Times of Zambia*, 13 August 1994; *Times of Zambia*, 16 August 1994, p.3; *Times of Zambia*, 11 October 1994). As a result, the council was owed over K3bn in water and sewerage charges at the end of June 1995 (KCC, 1995c).

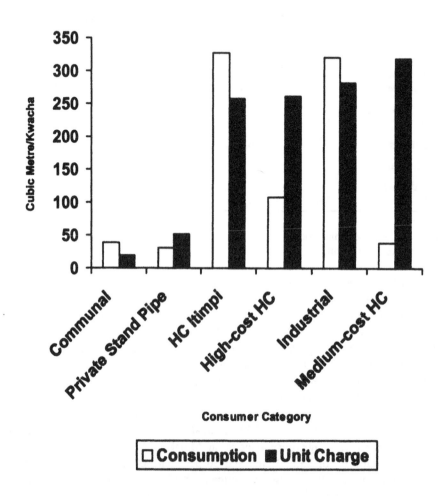

Consumer Category

□ **Consumption** ■ **Unit Charge**

Source: Derived from Kitwe Water Supply Project 1991, Table 3.3-1; Appendix Table 1 and 1994 tariff.

Figure 7.5 Monthly water consumption and unit charge

Assessment of Affordability Levels

One reason for low collection efficiency is the high default rate, which this section attempts to relate to incomes. Household income is defined as the aggregate income from all sources accruing to household members aged seven years and above (GRZ, 1991b). Housing expenditure on rent, water, electricity, candles, paraffin, charcoal, firewood etc. ranged between K4,000-K12,000 per month in 1993 (GRZ, 1993), of which Copperbelt residents spent an estimated two per cent on water (GRZ, 1991b). Household expenditure on water in 1993 should be between K80-K240 per month, yet in 1994 domestic water charges ranged from K800-K84,500 (Table 7.5 last column) per month, a wide disparity between water charges and available income. While water consumed from communal and private standpipes in low-cost housing areas is subsidized, metered consumption wherever it occurs (low-, medium- or high-cost areas) is not. Is it any wonder that in March 1995 KCC was owed over K3bn in water charges and had a collection efficiency of only four per cent (KCC, 1995a)? Although unmetered consumption from communal and private standpipes is subsidized, it is not necessarily affordable. Residents of Lusaka's George Compound (an upgraded informal settlement) continued using water from the boreholes they had dug themselves, even after piped communal supplies had been extended to them at a subsidized monthly rate, leaving the authorities threatening to backfill all boreholes in the compound so as to force residents to pay for the supply.

Cost Recovery - A Comparative Analysis

Kitwe's 1997 tariff structure is now compared with that of the Lusaka Water and Sewerage Company (LWSC). In January and May 1997, the Department of Water and Sewerage Services (DWSS) in Kitwe and LWSC respectively revised and published new water tariffs. While DWSS in Kitwe is physically and financially separate from the council (KCC, 1994b), it is part of the KCC organizational structure and the director attends council meetings. LWSC, on the other hand, was in the 1980s delinked from Lusaka City Council, and operates as an independent parastatal company. Table 7.6 depicts the level of consumption and associated metered charges using the Kitwe and Lusaka tariffs (this comparison is limited to metered consumption and excludes charges for communal and private standpipes in high-density areas).

A number of significant differences are apparent. Firstly, for domestic consumption of between 28 cubic metres and 48 cubic metres, LWSC charges are lower than Kitwe's, but as domestic consumption increases, the difference narrows (from about K2,000 for 28 cubic metres to a paltry K20 for 48 cubic metres).

Secondly, for domestic consumption of 109 cubic metres, 328 cubic metres, and for commercial and industrial consumption, LWSC charges exceed those for Kitwe. The actual change occurs when domestic consumption reaches 66 cubic metres; above that level, domestic consumption is charged by LWSC as if it were for industrial or commercial use, imposing a sanction for higher than normal domestic consumption. Compared with Behrens and Watson's (1996) normal consumption figure of 250 litres per person per day and a household size of 6.4 (giving a monthly volume of 48 cubic metres), the monthly allocation of 66 cubic metres is generous. While LWSC limits domestic consumption in this way, Kitwe's tariff is less punitive to high domestic consumption. While the highest consumption of 328 cubic metres is domestic, Kitwe's charges on this volume are lower than that on 322 cubic metres for industrial use, but LWSC charges are nearly the same.

Table 7.6   Comparison of 1997 water charges in Kitwe and Lusaka

Consumer Group	Estimated consumption (m³/month)	Kitwe tariff estimated charge (K/month)	LWSC tariff estimated charge (K/month)
Domestic	28	14,740	12,780
	34	16,060	14,880
	38	17,050	16,280
	48	19,800	19,780
	109	36,575	48,810
	328	96,800	268,360
Commercial	129	46,365	78,750
Industrial	322	129,580	276,700

Source: Author's calculations based on estimated Kitwe consumption from Appendix Table 4 and 1997 published tariffs in the Appendix.

Thirdly, although LWSC unit charges on the tariff rise with increasing consumption (Appendix Table 10) from K180 to K1,100 for domestic users, and K470 to K1,100 for commercial and industrial users, the use of standing charges (fixed charges levied irrespective of consumption) introduce economies of scale with a similar effect as the price discrimination noticed in the case of Kitwe (see earlier in this chapter). For example, when a standing charge of K4,000 is levied on all domestic consumption, a consumer who uses two cubic metres of water will be charged K4,360 (K4,000 fixed charge and K360 consumption charge). On the other hand, a consumer who uses one cubic metre of water will be levied K4,180 (K4,000 fixed charge and K180 consumption charge). The final unit charges for these two respective consumers being K2,180 and K4,180, lower for a consumption of two cubic metres and higher for one cubic metre.

Except for the blanket use of a standing charge, which rewards high domestic consumption (at least up to 66 cubic metres), the LWSC has shown more realism, creativity and consideration in structuring its 1997 tariff compared with

DWSS in Kitwe. Considering that LWSC is run as a semi-private commercial entity, while DWSS in Kitwe was at the time still a public enterprise, the analysis shows that private enterprise can behave in the interest of the community of low water users, and challenges the view that only public enterprises can behave equitably. Based on this analysis, Zambia's Water Sector Development goal (see earlier) is yet to be attained for Kitwe. A 100 per cent supply has not been achieved, the price structure is inequitable. The reference to competing user groups in the goal is of limited relevance because of the incidence of cross-subsidies. The huge arrears in unpaid water bills reflect unaffordable charges. As a result, continued water supplies on which the rate of recovery is very poor, are only being sustained at a great cost to the local authority.

Analysing the Impact of Different Pricing Policies

This section explores the economic and political viability of an alternative approach to water pricing, comparing the current pricing policy with one in which all consumers are charged on metered consumption, and also the marginal cost and average cost approaches.

All-meter approach: This assumes one common domestic tariff for all residential areas, whether low-cost or high-cost. Under the current system, except for those consumers with a meter (mostly the industrial/commercial users, high-cost residential areas and a few medium-cost houses), most domestic consumption in low-cost housing areas is charged on a fixed monthly rate, which according to the 1994 tariff was K2,000 for a private standpipe and K800 for a communal standpipe. For example, in Kawama site-and-service (see Appendix Table 4, row 10 column 5), 100 households had metered house connections, 340 households had un-metered private standpipes and 200 households relied on communal taps. Appendix Table 5 shows the situation as it would have been assuming that 1990 estimated consumption patterns remained unchanged in 1994, and that all households were charged on the 1994 tariff structure for metered consumption.

According to information supplied to the author (Water and Sewerage Questionnaire), KCC spending on water was K3.8bn in 1995. Assuming full production capacity of 83,000 cmd, the unit cost was K125/cmd; on reported production of about 55,000 cmd (KCC, 1994b), the unit cost was about K189/cmd. In either case, the fixed unit charges ranging between K21-59 (Appendix Table 4, last column) are lower than the unit supply cost of K125 or K189. It is reasonable therefore to assume that the difference between fixed charges and total costs for water supply is paid by metered consumers, whose unit charges range between K258-K322 (Appendix Table 4, last column), and represents a subsidy by consumers in high-cost housing and commercial/industrial users to low-cost housing built into the tariff (it has been used in the construction of Appendix Table 5 without adjustment). Although the current pricing policy (Appendix Table 4) and the all-meter approach (Appendix Table 5) make use of the tariff, the former combines metered and fixed charges and the latter reflects all-meter charges. If a subsidy element has been incorporated in the tariff to enable metered consumers to subsidize fixed charge consumers, then the use of the tariff without adjustment in the all-meter approach means that this subsidy is now being charged on all consumers. It seems reasonable therefore to deduce that the difference in potential annual revenue between K4.8bn in the current approach (K400.7 x 12, Appendix Table 4, column 6) and K7.8bn in the all-meter approach (K648 x 12, Appendix Table 5, column 6) reflects this subsidy. To analyse the economic and political merits of the all-meter approach in relation to the current approach, the subsidy element in the former is removed by applying a factor of 4.8/7.8 or 0.6154 to all unit charges in Appendix Table 5 to reflect metered consumption charges with no cross-subsidies. Appendix Table 6 compares current pricing policy (column 3) and all-meter (column 4) with a factor applied to all-meter charges to remove the subsidy element. It shows that 22,059 consumers (called losers in Appendix Table 6) currently subsidized would be negatively affected, paying an average of K196 per cubic metre from an average low fixed charge of about K40 per cubic metre, an increase of almost 400 per cent. 13,837 consumers (called gainers in Appendix Table 6) would benefit, unit charges dropping by about 40 per cent from an average high of K321 per cubic metre to K186 per cubic metre. Price discrimination in the tariff means that subsidized households (by definition expected to be poor, live in low-cost housing and with lower water consumption levels), still pay a high unit charge of K196 compared to K186 for others. The all-meter approach, by averaging distributional costs, introduces the allocative inefficiencies referred to later and with the present tariff structure exacts a higher unit charge from poorer households with low consumption patterns.

Marginal cost approach: Marginal cost is the increase in total cost resulting from a change in output of one unit (Parkin and King, 1992). The marginal element in this approach is related to the changes in total cost resulting from changes in distribution costs only. This approach is based on the argument that the pricing of public services to achieve allocative efficiency should reflect the variations in the cost of providing services to geographically disparate locations. Downing (1977) argues that the charging of a uniform average price to all users, without taking account of the variations in cost due to their unique locations, results in cross-subsidies. Appendix Table 7 attempts to apply the marginal cost pricing policy, using the assumed marginal production cost of K125 per cubic metre and a marginal distribution cost of K25 per cubic metre per kilometre. The figures are provisional and may have little relationship to the actual costs involved in the supply of water, yet they can identify the potential gainers and losers from an alternative pricing policy, and thus the political viability of a particular policy. Distances from the centroid of each area to the nearest reservoir have been scaled and used as representing the distance through which water has to be transported for each respective area. From Appendix Table 7, it can be seen that:

(i)   total marginal cost per year (K12.666m x 365 days) = K4.6bn, comprising:
(ii) total production costs per year (K8.998m x 365 days) = K3.3bn and
(iii) total distribution costs per year (K3.668m x 365 days) = K1.3bn.

The difference is small between the marginal cost approach (K4.6bn) and the estimated potential revenue from the current approach (K4.8bn, in Appendix Table 4). The assumed marginal production and distribution costs are little different from the combined real costs in 1994, assuming that the tariff used in calculating potential revenue attempts to recoup these. A departure to marginal cost, although economically the most efficient in terms of allocative efficiency, would lead to the greatest cost to those on the fringe of the city, mainly low-cost domestic consumers in site-and-services and informal settlements (e.g. Luangwa and Mwaiseni respectively). It is the most equitable and efficient approach because consumers pay the actual cost of distribution and production. The alternative to the marginal cost approach is called average cost, which is now discussed.

Total average cost approach: Total average cost is the sum of all the average costs in production and distribution (Parkin and King, 1992). This approach matches the charges to the average total costs involved in supplying water. The bottom row of Appendix Table 7 shows the average cost of distribution as K51 per cubic metre and that of production K125 per cubic metre, giving an average total cost of K176 per cubic metre. There are obviously other costs involved in supplying water, but only these two have been considered here. Although cited as an alternative to marginal cost pricing, this approach can be criticized for its allocative inefficiencies by way of cross-subsidies, because it fails to account for the increased costs of distribution beyond that of the average cost (Downing, 1977). Figure 7.6 illustrates this incidence of cross-subsidies under average cost pricing.

OP1 is the marginal cost of production at source and P1B is the curve representing marginal distribution costs at various distances from O. Although all consumers pay the average cost P2, for those located between O and D1, this is higher than the marginal cost of production and distribution. On the other hand, for those beyond D1, the average cost P2 represents a price lower than the marginal cost.

In Kitwe, industrial consumers and nine residential areas (among them six low-cost housing areas, representing a client population of 13,772) would, under the average cost approach, pay more per unit, potentially subsidizing consumption in the CBD and nine outlying residential areas with a combined consumer population of 22,124. It is worth pointing out, however, that the subsidized residential areas under average cost are eight low-cost housing areas, largely informal settlements and site-and-service schemes.

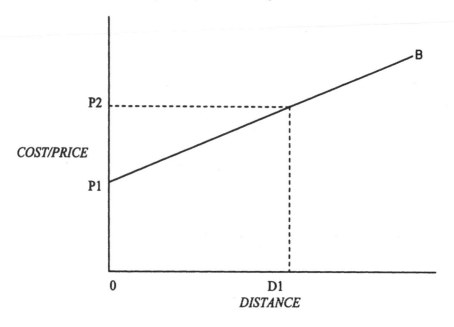

Source: Author.

Figure 7.6 Incidence of cross-subsidies in average cost pricing

A synthesis: The overall identity of consumers benefiting or losing from different pricing approaches does not change, because the current subsidy in the existing pricing policy nullifies the price differentials between different approaches. A change from the current approach to any of the three other alternatives (all-meter, marginal and average cost) would adversely affect about 22,000 consumers.

The total average cost policy seems to offer the greatest promise for Kitwe's present land use pattern and water distribution network, having the potential to recoup the actual costs of supplying water. While it subsidizes those consumers located beyond the radius defining the average cost (Figure 7.6) and hence inequitable to those consumers located up to this radius, Kitwe's land use pattern shows these outlying areas as mostly poor, low consumption clients for whom it might be considered morally right to subsidize. The total average cost is also much simpler to administer, and would not raise the same issues as the marginal cost approach. All consumers would, under the average cost policy, be expected to pay a uniform unit charge, the final bill depending on the volume consumed. There would therefore be no question of consumers arguing 'Water is water, why should we be charged more per unit than others?'

In practice the situation is less straightforward, because some consumers within the average cost distance, for whom it is argued that it is morally right that they subsidize outlying poor areas, will themselves need to be subsidized, for example, the planned low-cost council housing areas of Buchi and Kamitondo

(Figure 5.11). Others outside this distance although generally referred to as poor may not be, and will not therefore need the subsidy which the use of average cost automatically gives to consumers outside the average cost distance.

Because levels of consumption and income differ from one group to another, the inherited segregated structure of Nkana-Kitwe could be put to good use, as these cellular units form fairly homogenous social groups in which consumption does not differ markedly from one consumer to another. These well-defined residential units could be used to assess the average volume of water used, allowing basic need and subsidies to be determined more accurately.

The alternative policy proposal should, therefore, strive to meet the consumers' basic need for water. In practice, a household's need for water would be assessed and compared against the available expendable income to meet this need. Where this income falls short, a subsidy should be given up to the amount necessary to pay for the basic need. Any consumption over and above this basic need has to be charged normally based on the average cost rule. It should in theory be possible to use cross-subsidies to meet the basic need. In order to increase affordability and therefore reduce the level and negative economic effects of subsidies, the basic-need consumption should be charged at cost price with no profit mark-up. The new tariff structure will reflect a lower unit cost for the first few units of consumption (representing the basic need), and then a higher one thereafter. This needs-based tariff proposal contrasts with the current one, which rewards high consumption and does not limit the amount of subsidized consumption.

Conclusion

Observations about the distributive function of urban management and the need for moral principles to guide allocative decisions were made in Chapter 1. This chapter has analysed aspects of equity and morality in the distribution and pricing of water. An eclectic approach combining the institutional and interpretive methodology supported with empirical data has been used.

In discussing the institutional structure of water supply in Nkana-Kitwe, both the mining company and the local authority own water treatment plants and supply the city with water. The notion of social overhead capital was advanced as the basis for the mining company to extend its water supplies to its township, and later to the public township. We have seen in the case of water how the fulfilment of the company's self-interest is linked to greater social concerns.

It has been estimated that about 91 per cent of Nkana-Kitwe's population has access to piped water, with higher service options such as house connections mainly restricted to high-cost residential areas. Constraints in production and distribution have been identified with mechanical and structural problems due to the age of installations and public abuse of infrastructure.

Consumption increased rapidly between 1955 and 1974. After 1974, although the city expanded, supplies remained limited and consumption levels in 1990 have been estimated to fall below their 1974 levels. In relation to what is considered as normal consumption, the city of Kitwe has high domestic water consumption, which, if moderated, would supply an additional one and a half times the number of existing domestic consumers. But, a comparison of income against water charges shows that, even if consumption was minimized and there was now water to supply the rest of the city, there would still be the problem of affordability.

Water proposals in the City of Kitwe Development Plan have been criticized for over-estimating both future consumption and population, not acknowledging the initial deficit in service coverage, and their inability to account for the city's financial constraints.

The second part of this chapter has concentrated on the financial aspects of water supply, examining issues of cost recovery and alternative pricing policies as they relate to equity and moral acceptability.

The council's objective in fixing water charges is to match the overall expected revenue with total expenditure, while cross-subsidizing to cover the cost of consumption in low-cost residential areas. As long as consumption in these areas is subsidized, the fixed water charges levied on these households falls below cost price. But, if these households were levied on the basis of metered consumption with no subsidies, the main finding has been that the water tariff is weighted in favour of high consumption by high-cost residential areas who pay a lower unit-charge. This common economic approach is challenged on the basis of the potential immorality of exacting a high unit charge from the poor in low-cost housing, and a lower one for the relatively rich in high-cost residential areas. It has been suggested that subsidized consumption should be limited to the basic need, thus giving relief to those metered households who, in most cases, are not able to pay both for their own consumption and to subsidize others.

A further observation made under cost recovery is the grouping of commercial and industrial consumers under the same sub-tariff structure, which is argued to be uneconomical because it allocates a disproportionately high initial volume to commercial users who consume a comparatively low amount of water.

To the extent that cost recovery is a problem, a number of contributory factors have been identified, among them: historically low salary structures which leave little expendable income to pay for such services (designed for a high-income European population) leading to high default rates; historically low tariff structures reflecting the colonial non-profit objective for services to African areas in which water charges bear no relationship to the real cost; and the council's own system inefficiencies.

A comparison between Kitwe's DWSS tariff and Lusaka's LWSC shows that the latter charges less for low consumption than DWSS, and that LWSC attempts to limit the level of domestic consumption by imposing higher rates for excessive domestic consumption. Although LWSC has shown consideration for low consumption and innovation in limiting domestic consumption, it also fails to deal with the problem of discrimination based on consumption. The evidence of comparatively lower domestic charges for low domestic consumption under the LWSC attests to the possibility of a parastatal company pursuing moral objectives with regard to the domestic consumption of households in low-cost residential areas who are relatively low consumers of water.

Different pricing policies for Kitwe's water distribution, settlement pattern and consumption levels were analysed. With the all-meter approach compared with the current approach in 1994, a consumer population of about 14,000 in high-cost residential areas, commercial and industrial areas subsidize the consumption of about 22,000 households in low-cost housing, a ratio of 1:1.6. This group of subsidized households would be adversely affected if the subsidy were removed and all consumers were expected to pay the full cost of metered consumption. In examining the marginal cost approach, it has been implicitly argued that the current pricing approach is inefficient in its allocation of water charges, because water charges are uniformly set according to the consumer group (commercial/industrial, domestic house connection, communal and standpipe), taking no account of relative location and corresponding differential cost of supply. The main objective in fixing water charges has been identified as balancing the total expenditure and total income. Except for low-cost domestic consumers, who enjoy a fixed subsidized rate, other consumers pay according to the level of metered consumption in their group. Whereas economists in developed countries have argued for marginal cost pricing because of allocative efficiencies only, analysis of the functional land uses in Nkana-Kitwe shows that informal housing areas have located on the fringes of the city, and marginal cost pricing policy would hurt these people the most. An alternative approach balances questions of equity, affordability and willingness to pay against other fiscal objectives. Although criticized for its allocative inefficiencies in comparison to the marginal cost approach, the analysis has shown that the total average cost, given Kitwe's water distribution and settlement pattern is preferable, because it has the potential not only for cost recovery, but for cross-subsidy to poor households on the periphery of the city.

Because of the level of subsidy in the current approach, the overall identity of the consumer population affected by a change in pricing policy remains the same regardless of differences between these policies. The level of subsidy is considered too high, but morally unacceptable and politically risky to withdraw at once. Urban management in this situation has to balance its technical functions (among them the ensuring of allocative efficiency) with its moral responsibility of allocating this basic human need to competing users, among whom are the poor and hence vulnerable groups.

If KCC ignored the moral question of exacting the highest charges from the poor on the periphery of the city, the marginal cost approach is the most (economically speaking) equitable, because it charges each group of consumers the actual cost of supply. Moral considerations suggest that an approach offering the highest benefit to the least advantaged, in this case the average cost pricing rule, is adopted. The current policy already offers a substantial subsidy to the poor, but does not take account of the ability of one group of consumers to subsidize another. For example, a consumer is automatically charged a higher rate because they have a private meter, even though they may be in the low-cost area. The tariff as currently constituted does not limit subsidized consumption, and is designed to discriminate low metered consumption for higher charges. This can be regarded as immoral because the poor are also generally low consumers. An alternative which addresses both the economic need for sustaining the service and the social objective of meeting the basic need for water at cost price is necessary. Although advocating the total average cost approach, the subsidy element has to be re-examined and re-structured to fit into an alternative pricing policy which addresses both economic and moral issues.

Chapter 8

The Form and Agency of Housing Provision and the Nature of the Housing Problem in Nkana-Kitwe

Introduction

The nature of housing form (one of the most common forms of material culture) and problems are uniquely influenced by the social matrix (political, economic and religious organization) of the community. Given the dynamic nature of society, it could be argued that housing is always changing with the changing political, economic, religious organization and social mix of the population. Two aspects predominate approaches to the study of housing, these are social and economic. In this chapter, certain social and economic aspects of housing are considered.

The first aspect is the form and agency of housing provision in Nkana-Kitwe, changes over time and the financial structure of local authority housing. Returning to two questions posed in Chapter 6 (why some people are not able to acquire land through the formal process and why the homeownership concept has been difficult to realize), the second section examines the level of public housing supply and demand, and the relationship between the cost of land, housing and finance in Kitwe.

The third section addresses the emergence and progress of informal housing, exploring official responses to resettle these people in site-and-service areas, and the relative level of provision between the local authority and informal sector.

A broad assessment of housing need in Nkana-Kitwe is made in section four, followed by practical proposals for improving access to land and housing, including a recommendation to densify existing settlements supported by a detailed analysis and comparison between specified and actual building standards.

Secondary data sources and primary empirical data from the archives, government statistical sources, valuation rolls and local authority reports provide the main sources of data for this chapter.

The Agency and Form of Formal Housing

Except for Tipple (1978, 1981), little has been written on the agency and form of housing in Nkana-Kitwe, and pre-dates the re-introduction of plural politics (1990) and the decision to sell all public institutional housing (1994). This section contributes to existing research by: exploring more fully, and bringing up-to-date, the complex factors influencing the provision of housing; exploring the changes through time of the built form of housing; and examining the financial structure for local authority housing. The section is organized and discussed under different identified agencies, against the development of national housing policy discussed in Chapter 4.

Mine Housing

Improvements to African housing in the Copperbelt towns in the 20 years before independence were part of a debate around the so-called 'stabilization' of labour (Chapter 4). The Territorial government's file on the subject (ZNA: SEC 1/1320) defined the stabilization issue as 'whether efforts shall be made to build up a permanent mining population, or whether the miners shall be encouraged to regard the village as their base and spend short periods in the mines'. As early as 1930, a visiting German anthropologist advised that creating stable communities of permanent African residents would provide 'an efficient and economic stock of native labour' (cited in Meebelo, 1986, p.166). Although initially sympathetic to the stabilization of labour, after the Copperbelt labour unrests of 1935, the mining companies preferred a turnover of African labour, seeking to avoid unionization and the establishment of family life in the towns: 'detribalisation and urbanisation should not be encouraged in any manner' (RST Chairman in 1937, cited in Meebelo, 1986, p.166). Various representations, for example, from the London-based Aborigines Protection Society and parliamentary questions asked by Arthur Creech-Jones (later Colonial Secretary in the post-war Attlee administration) in 1943 on the need for family housing for the Africans on the Copperbelt, led to a meeting in Kitwe of Territorial officials in 1944 to lay down rules for stabilization. After 1945, mining companies in Northern Rhodesia accepted the concept of a stable, permanent labour force, and, in an effort to forestall possible political agitation because of urban problems created by migration and unemployment, it was decided to give African workers a measure of material advancement (Quick, 1975). We now proceed to discuss and illustrate the progressive improvements in housing - particularly African mine-worker housing developed by the mining company and its subsidiaries both within and outside the mining township.

Although Nkana was declared a mine township in 1935, the mine itself opened in 1923 and production commenced in 1932 (Copperbelt Development Plan, n.d). Some of the earliest European residential quarters were of Kimberley bricks and had thatched roofs, as did the temporary compound built to accommodate hundreds of African workers (Figures 8.1 and 8.2).

The European section (Nkana West) grew from 64 residences in April 1930 to 166 in the succeeding 15 months, complete with a hospital and recreation hall, all provided with water, sewerage, electricity and roads (Bancroft, 1961).

Source: Bancroft, 1961.

Figure 8.1 Part of the European township at Nkana Mine in 1929

Source: Bancroft, 1961.

Figure 8.2 African compound at Nkana Mine in 1929

As prospecting continued and the mine became more viable, temporary settlements were re-developed to assume a more permanent form. By 1950, all old housing in Nkana West had been re-developed with more permanent building materials (City of Kitwe Development Plan, Survey of Existing Conditions). From about 1940, the development of European mine housing spilled outside the mine township and crossed the Kitwe-Ndola trunk road to create Nkana East.

As for African housing, the re-development derived from the barrack-type housing which offered ease of regulation (Home, 1993, 1997, 1998; Demissie, 1998). The site was cleared of all trees and no hedges above a certain height were allowed. Each block accommodated ten families, each family with one private room and a kitchen, toilet facilities and water were shared between families in one or more blocks (Figure 8.3). A compound manager, appointed under the Mine Township Ordinance (Chapter 4), supervised the compound. Referring to the regimental character of African mine housing, Kaluba (1994) reminisces:

> Heavy sleepers were woken by the personnel officer nicknamed 'Changachanga' who went around the mining township shouting in Chibemba 'Mukambatuke! Mukambatuke!' (Get off her! Get off her!) while beating a tin pan.

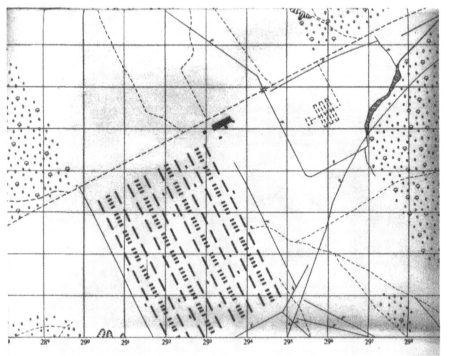

Source: Zambia Survey Department, 1:5,000 Topo-Cadastral Series Map, Sheet No. 2880.

Figure 8.3 Layout of barrack-type housing in Wusakili, Nkana, 1959

Although the government opposed this form of housing, preferring single huts, the mining company claimed that the 'natives... are comfortably housed and seem to be quite satisfied and contented' - only a month before the 1935 strikes revealed the scale of African mine-workers' discontent (ZNA: SEC 1/1535, Vol. II). The inquiry into the 1935 strike criticized at Nkana 'the barrack type in long rows, all shade trees having been cut down... an impression of overcrowding, and lack of privacy... austere quarters rather than houses' (Northern Rhodesia, 1935, pp.62-3 and 140). The inquiry preferred the paternalistic arrangements at Luanshya, with trees, fences, outbuildings and the keeping of livestock allowed:

> ...the object at that mine is to maintain the tribal and village customs of the natives as far as is possible in the conditions... They become quite proud of their little homes (Northern Rhodesia, 1935, pp.57-8).

Mine housing, according to recommendations of this inquiry, needed to perform according to Rapoport and Hardie (1991) the prosthetic function of helping people to settle into the new industrial environment.

The 1935 strike gave the mining company a severe fright and contributed to a change in the mine's housing policy towards detached bachelor dwellings. The old Wusakili township illustrates the form of these bachelor dwellings: mud brick walls rendered with cement, each dwelling enclosed by a hedge (kept below three feet in height, for ease of external surveillance), no windows, but only ventilation openings beneath the corrugated-iron roofs (Figures 8.4a/b and 8.5). Internally two rooms provided a total floor space of under 200 square feet (18 square metres) and a typical plot was about 100 square metres. Washing and toilet facilities were centrally provided to cater for a number of households.

While this type of housing was an improvement over the grim barrack type, the government still judged it as falling below accepted standards of ordinary decency. Commenting on such squalor, the Brown Commission observed:

> Houses with one habitable room, without windows or separate ablutions, should never have been built at all and it is important to realise that it is moral and not building standards which have improved with the passage of years (GRZ, 1966c, cited in Turok, 1979, p.13).

From 1939 to 1944, the mining company built 2,797 dwellings for Africans in Nkana, which represented 83.5 per cent of the total new dwellings (the rest being 93 for central government, 493 for local government and 48 for the railways) (PRO: CO/795/133/45384). Further improvements to African housing were made during the late 1940s and early 1950s: new dwellings offered increased floor areas of 25-55 square metres, with up to three habitable rooms, individual toilets and washing facilities.

That Africans had come to stay in urban areas is evidenced by the fact that the average length of stay at the mines for the 1959 labour force was about 63 months compared to the average of 13 months since mining began at Nkana (Anon, dated about 1959). In addition, more than 60 per cent of the married African labour force had their families with them. This level of 'stabilization' was matched by the provision in the early 1950s of 60 per cent family housing units (Hailey, 1957, p.576), and the provision of primary schools (Tipple, 1981).

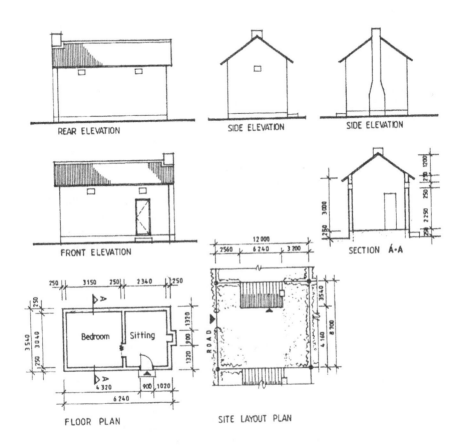

Note: All dimensions in mm.
Source: Wetter, M., Copperbelt University, Kitwe, Zambia. Personal communication.

Figure 8.4a Architectural drawing of old low-cost housing in Wusakili, Nkana

Source: Author, 1997.

Figure 8.4b Old low-cost housing in Wusakili, Nkana

Source: Author, 1997.

Figure 8.5 Communal washing and toilet facility in Wusakili, Nkana

It is estimated that by 1959, there were 8,911 houses in the African townships of Nkana comprising single and married quarters with two bedrooms, a living room, a kitchen, a storeroom or pantry, nearly half with an internal toilet and shower, and all built of permanent material (Table 8.1). Rents ranged from 7s 6d to £2 17s 6d, and these were regarded as sub-economic (Anon, dated about 1959). Commenting on life in African mine townships, a mining company brochure noted that:

> The European 'look' of the mine townships with its rows of neat, orderly homes is deceptive. Underneath, the way of life is fundamentally African. Dishes are left to dry in the sun and the little yards separating each home are like miniature frontages of kraals with their cooking pots and piles of firewood... Fowls dart across the roads under the flying feet of children. Mothers bake in the sunshine when the household chores are done... hasten with their babies strapped to their backs... wander in chattering groups... (Anon, dated about 1959, p.21).

African housing improved in structure and space (Figure 8.6 and Table 8.1) compared to the austere and cramped conditions of the barracks, but also helped Africans settle into the new urban areas by promoting a lifestyle similar to village life. Table 8.1 is a summary of mine low-cost housing built before independence in 1964.

Source: Author, 1997.

Figure 8.6 Miners' cottages in Chamboli's Natolo Street, Nkana

By 1964, the building of low-cost houses by the mines had ended. After independence, mine housing was mostly built to medium-cost standards; for example, Miseshi and Cha-Cha-Cha - the latter named after the dance, which in Zambia is associated with the stormy period of pre-independence agitation. Although the mining company, in the 1970s, expressed an intention to demolish 5,000 sub-standard houses at Wusakili and Mindola, and replace them with larger houses at lower densities (KCC, 1976), in practice this did not happen.

Table 8.1   Mine low-cost housing at Nkana

Period of construction	Number of dwellings built to 1972	Number of habitable rooms	Approx. internal floor area (m^2)	Facilities
Pre-1940	5,234	2	20	Communal showers and lavatories
1952 onwards	1,058	3	25	Shower and lavatory for each house, attached or in out-house
	1,472	3	36	As above
	743	3	50	Internal shower and lavatory, separate or in same room
	635	3	40	Shower and lavatory in same room, internal or attached to houses
	400	3	48	Internal shower and lavatory in same room
	184	3	55	Separate internal shower and lavatory

Source: Tipple, 1978, p.157.

In terms of nett densities, mine housing was developed to an average of 40, 27 and five dwellings per hectare for low-, medium- and high-cost houses respectively. Figures 8.7 and 8.8 show the contrast between low-cost (high-density) and high-cost (low-density) mine housing.

To summarize, the pattern of African mine-worker housing in Nkana shows the transition to stabilization, reflected in the shift from bachelor to family units. The history of African housing can be physically traced on the ground through the successive 'townships', where gradually housing standards (although far short of those provided to European residents) were improved. Housing provision was politicized, and dominated by tensions between mining capitalism and a paternalist but weak government. If the negotiations over the twin townships saw the mining company pushing a slow moving government (Chapter 5), in the matter of African housing improvement it was the other way round, the government prodding a reluctant mining interest to invest in social infrastructure. The next section examines the government's own record of providing housing in the public township.

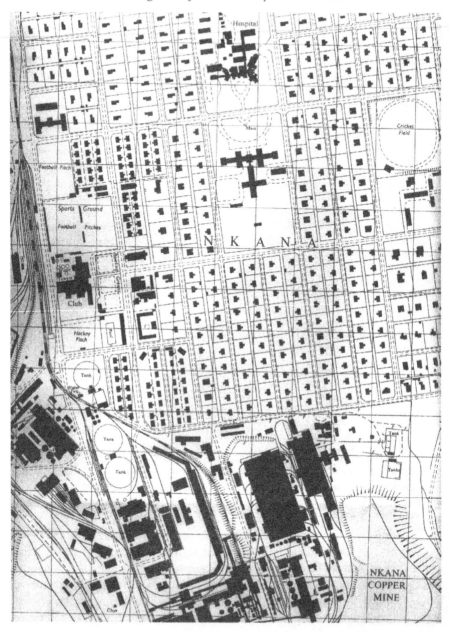

Note: 1 square = 100m x 100m.

Source: Zambia Survey Department, 1:5,000 Topo-Cadastral Series Map, Sheet No. 2880.

Figure 8.7 Mine low-density housing in Nkana West

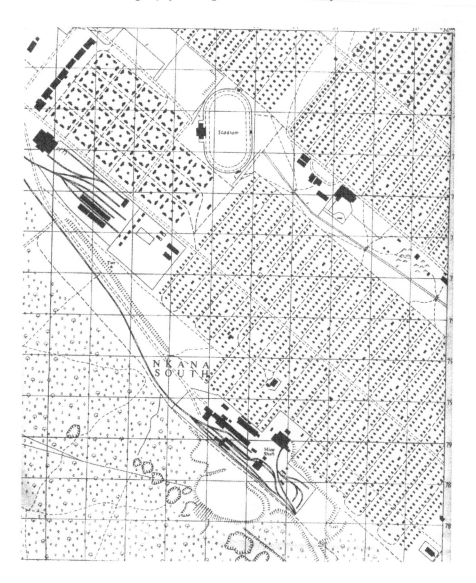

Note: 1 square = 100m x 100m.

Source: Zambia Survey Department, 1:5,000 Topo-Cadastral Series Map, Sheet No. 2878.

Figure 8.8　Mine high-density housing in Wusakili, Nkana

Local Authority Housing

The rapid development of Nkana Mine in the 1930s required the government to establish a public township to serve both the needs of the mine population and others, but financial constraints could not allow for such public investment. In a co-operative effort between the mining company and the government's Provincial Commissioner, a number of European residential and commercial plots and an African trading and residential area were proposed for the public township. The type of housing suggested by the mining company was 'the regulation compound as at present erected at Nkana' (ZNA: SEC 1/1535, Vol. II), arguing that single huts were too expensive. In a minute to the Governor, the Provincial Commissioner, Goodall, let his distaste for the type of housing suggested by the mining company known: 'I observe that they propose to build in the barrack style to save expense. My view of this type of accommodation is that it means nothing less than the creation of native slums' (ZNA: SEC 1/1535, Vol. II).

After detailed spatial and financial comparisons with Lusaka's Hodgson type of round huts (ZNA: SEC 1/1535, Vol. II), Kitwe Location, an African sector of the public township, was built in 1942; the houses were of poor sun-dried brick and had a rectangular lean-to kitchen. In all, 493 units and five sanitary blocks (in all costing £21,000) were built, compared to 2,797 built by the mining company during the same period (PRO: CO/795/133/45384).

Source: Author (sitting in foreground), c.1997.

Figure 8.9 Lusaka Hodgson-type hut

By 1950, the African population in the public township was concentrated in two temporary and unsightly settlements each housing 400 families. The first, called KMB (after Kitwe Management Board) was situated west of the present Nyerere road and south of Kitwe stream. The second was sited between Chibuluma road and Kitwe stream east of Nyerere road (City of Kitwe, 1970).

The first permanent post-war low-cost local authority housing programme was initiated in 1950 in Buchi (Figure 8.10) to re-house families from the two temporary settlements above.

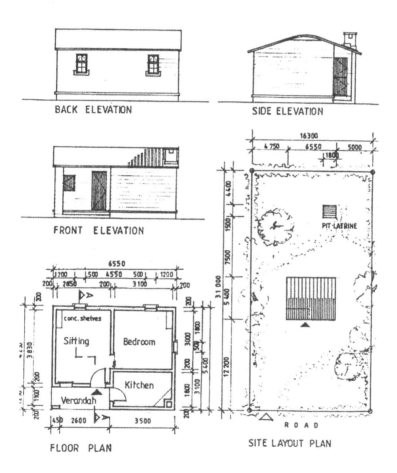

Note: All dimensions in mm.
Source: Wetter, Copperbelt University, Kitwe, Zambia. Personal communication.

Figure 8.10   Architectural drawing of a low-cost local authority house in Buchi, Kitwe

Because of the rapidly growing African population, the re-housing programme soon led to the further development of Kamitondo, Kwacha and Chimwemwe in quick succession, with Chimwemwe (meaning happiness) being the largest single local authority housing township in Kitwe (Figures 8.11 and 8.12).

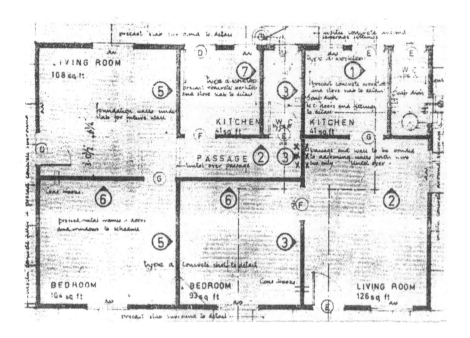

Note: The wall marked XXX could later be removed to convert the building into one house.
Source: Municipality of Kitwe, 1962.

Figure 8.11   Local authority housing in Chimwemwe, Kitwe

Demand for low-cost housing kept on rising. At the end of March 1961, the demand for married quarters was over 1,000 - a year described as one of 'consolidation and improvement...with a fair amount of expansion' in which only 499 houses of the projected 1,100 were completed in Chimwemwe (Municipality of Kitwe, 1961, Section 9). Meanwhile, a similar shortage of European housing, due to restricted access to mortgage finance, led to the local authority lending £50,000 to the building societies with the condition that 80 per cent of the money was made available to persons wishing to build in Kitwe. As a result, a large number of people were enabled to borrow and build their own houses (Retrospect, 1962).

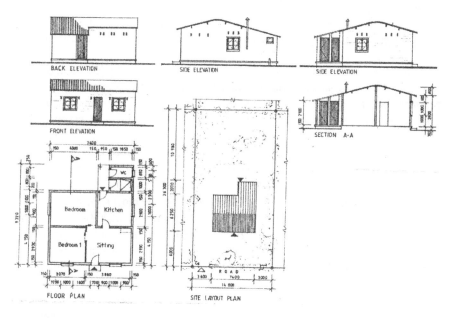

Note: All dimensions in mm.
Source: Wetter, Copperbelt University, Kitwe, Zambia. Personal communication.

Figure 8.12   Architectural drawing of local authority housing in Kwacha, Kitwe

By 1963, the African housing problem reached acute proportions and led to the tabling of a report meant to:

(i)   review the council's attempts over the years to keep pace with development;
(ii)   examine its achievements in this direction;
(iii)   analyse the recent phenomenal growth of the problem; and
(iv)   illustrate the council's dilemma in its efforts to meet the demand for housing.

Besides the continuing Chimwemwe housing programme, no significant proposals for extra housing were forthcoming, only plans to improve standards for a limited number of new rental housing and existing stock. At independence in 1964, local authority housing stock had increased from 1,141 in 1953 to about 10,000, while the council house waiting list was estimated at 5,000. Except for Chimwemwe, with a density of 47 dwellings per hectare and Buchi's 18 dwellings per hectare, nett densities in the public township averaged 31, 16 and six dwellings per hectare for low-, medium- and high-cost housing areas respectively. The reason for the higher than normal density in Chimwemwe is attributed to a later sub-division of some 600 houses. Buchi exhibits a very generous density for low-cost housing because of the allowance for pit latrines. Figures 8.13 and 8.14 illustrate the contrast between a high-density and low-density area in the public township of Kitwe.

Because of the obligation placed on the employer and the local authority by the Urban African Housing Ordinance of 1948 (see Chapter 4) to provide or cause the provision of accommodation to all African employees, local authority housing was mostly employment-tied. Not only was the local authority compelled to provide housing for African employees, it was happy to adopt this approach as policy in house allocation to attract investment. Council housing policy in 1970 favoured the housing of key personnel for new investors and for those wishing to expand existing investment (City of Kitwe, 1970). Although the incidence of 'self rent-paying' tenants (as opposed to that paid by the employer on behalf of the employee) increased to about 1,000 in 1962 (Municipality of Kitwe, 1962), the predominance of tied-housing necessarily excluded those not employed or without a steady income necessary for regular monthly rentals. Of the total male population of 58,300 in Nkana, Kitwe and Garneton, 55 per cent or 31,870 were unemployed and excluded from employment-tied housing (Retrospect, 1963). The effect of this housing problem will be discussed when examining the significance of squatter housing. Rentals for council housing in 1961 were fixed at £2 16 0d per month for married quarters and £1 8s 6d for single quarters (Municipality of Kitwe, 1961).

In addition to being tied to employment, the allocation of public housing, especially after independence, has been riddled with politics. In reference to political interference in housing administration from councillors, a report noted that it is hard to allocate houses according to procedure (KCC, 1993/94), in addition to councillors facilitating re-entries of tenants evicted for non-payment of rent (KCC, 1995d). This conflict came to a head when a motion was moved by a councillor to overhaul the Department of Housing and Social Services for incompetence and corrupt practice (KCC, 1995b). These constant wrangles between the politicians and administrators reveal the vulnerability of a public allocation of land and housing used for self-enrichment or political patronage.

Restoration of freedom of movement and residence to all Zambians at independence in 1964 led to a wave of migrants from rural to urban areas. By 1967, housing in Kitwe had become a major problem. The council's waiting list had risen from 5,000 in 1964 to 12,965 in 1967, while the four-year housing fund proposals could only provide for 1,630 houses in Ndeke township and 858 site-and-service plots in Kwacha East (City of Kitwe, dated about 1967).

The SNDP (GRZ, 1971) put an end to public rental housing, and instead provided for home ownership on fully serviced sites provided by the council (Chapter 4). Under this scheme a number of houses were built and sold by the council in Ndeke (Figure 8.15), Kwacha East and Riverside (KDC, 1981, 1983, 1985). Although the scheme proved popular to those who could afford the houses, low-income groups who could not afford the cost wished that the council could build rental units (KDC, 1983), a view shared by council officials because the scheme did not benefit low-income clients (KDC, 1981). As a result, the council's waiting list kept on rising. Empowered to build rental houses again in 1975 (Tipple, 1976b), the council entered into negotiations to raise K7m to build 500 low-cost and 100 medium-cost houses for rent (KDC, 1983). In 1985 the council was building 150 medium-cost rental houses and another 100 low-cost rental houses in Ndeke G, with an intention to build more (KDC, 1985).

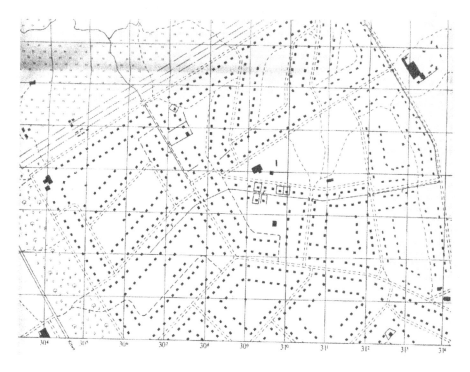

Note: 1 square = 100m x 100m.
Source: Zambia Survey Department, 1:5,000 Topo-Cadastral Series Map, Sheet No. 2886.

Figure 8.13 High-density housing in Buchi, Kitwe

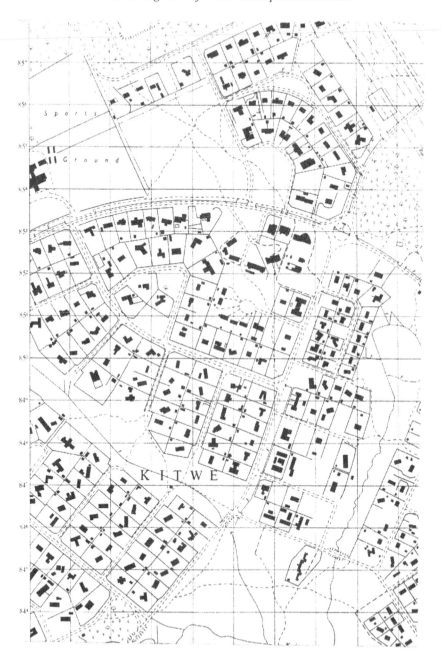

Note: 1 square = 100m x 100m.
Source: Zambia Survey Department, 1:5,000 Topo-Cadastral Series' Map, Sheet No. 3284.

Figure 8.14 Low-density housing in Parklands, Kitwe

Table 8.2 Local authority low-cost housing in Kitwe, 1951-62

Date	Compound	Single quarters	Number of rooms				Facilities
			2	3	4	5	
1951-1953	Buchi	--	1,085	56	--	--	Pit-latrine and outdoor water supply to each house
1953-1954	Kamitondo	112	114	206	--	--	As above
1955-1956	Kwacha	--	368	213	--	--	Pit-latrine and internal water supply
1957-1958	Kwacha	--	--	1,482	--	--	As above
1958-1959	Chimwemwe	--	--	750	--	--	Internal shower and lavatory
1959-1961	Chimwemwe	--	--	1,100	--	--	As above
1961-1962	Chimwemwe	--	1,200*	--	2,300	14	As above

Note: * Designed as 600 four-roomed detached dwellings, but converted to 2 x 2 semi-detached dwellings.

Source: Tipple, 1978, p.159.

Source: City of Kitwe Development Plan, Survey of Existing Conditions, p.72.

Figure 8.15 Local authority Ndeke township, Kitwe

The demand for housing shortly after independence, was in both low-cost and high-cost areas. While 133 serviced high-cost plots in Riverside remained unsold in 1966, the initiative to develop a cluster of 30 houses by a building firm led to all other vacant plots being taken up. By 1967, the council's plans to service a second area of Riverside had to be modified: plots originally of half an acre were reduced to between one-third and one-quarter of an acre to meet demand, for which a waiting list had already reached 130 (City of Kitwe, dated about 1967). As for the financial structure, a segregated accounting system provided for the self-financing of low-cost housing (Table 8.3).

Table 8.3   Percentage revenue and expenditure in Kitwe, 1960-64

| | Revenue | | | Expenditure | | |
Year	Low-cost[a]	High-cost[b]	Gov't grant	Low-cost	High-cost	Reserve fund
1960	48.0	47.0	5.0	49.0	51.0	--
1961	42.6	52.8	4.6	41.0	51.0	8.0
1962	44.2	51.0	4.8	44.0	46.0	10.0
1963	47.3	48.3	4.4	47.0	47.0	6.0
1964	49.6	46.6	4.0	50.0	44.0	6.0

Notes:
a. = African housing rents and beer sales
b. = Electricity, water, rates and other charges

Source: Compiled by author from various local authority reports.

Table 8.3 shows that local authority revenue from low-cost housing areas between 1960-64, consisting of income from rent and beer sales, was spent on the maintenance of housing including roads, sewers and water. About three per cent of the income from beer sales was spent on community or welfare services in the low-cost areas with a larger proportion re-invested in liquor-undertaking and a small transfer to the reserve fund. Revenue from high-cost areas came from electricity, water, general rates and other charges. A government grant averaging about five per cent of the total revenue was the only external source in the revenue account.

An examination of revenue and expenditure levels in the low-cost and high-cost areas shows a near match between income derived and expended in each area. A striking revelation concerning the level of revenue derived from the low-cost African and high-cost European areas is that it is almost in the ratio of 1:1 compared to the national income ratio of 1:9 in 1964 (Table 5.4). Before this is hastily interpreted as constituting a heavy charge on low-income African households, a comparison of population levels in 1962 shows 1,200 residents in high-cost areas and 41,000 in low-cost areas (Retrospect, 1962; Municipality of Kitwe, 1962). A comparison of *per capita* levels of revenue and expenditure, shows that low-cost areas benefited from the effects of cost sharing. It is also clear from the revenue and expenditure ratio of 1:1 that if the same money was received and expended to two significantly different population levels, then either the standard of services was poor in one group, or if services were of the same standard, then the larger population was being subsidized. While a cross-subsidy element cannot be discounted, especially after 1966 ('one town concept' - Chapter 6), it has been shown in the case of water (Chapter 7) that the standard of services is comparatively lower in low-cost areas.

Until 1963, the local authority's housing revenue account was always in credit. However, in 1964 a deficit of £30,325 was recorded, rising to £59,100 and £67,000 in the subsequent two years. If these deficits were expressed in terms of rental increases, they translated to increases of 10 shillings per month per household in 1964 and 11 shillings per month per household in 1966 on rentals which had not been increased since 1958 (*Kitwe Town Crier*, 1965a, 1965b). Under pressure in 1965 to increase rental charges as a condition for getting a government grant of £500,000 for more houses in Chimwemwe, but mindful of the impact such a sudden increase would cause to households, the local authority decided to meet these deficits from the surplus out of beer sales and a proportion of personal levy proceeds. The transfer to the housing account of beer sales profits meant for capital improvements to low-cost housing areas meant that such improvements could not be implemented. After a recommendation to the government made by Kitwe's municipal council in 1965 for a generous housing subsidy to the lowest paid, so as to keep rentals at reasonable levels (*Kitwe Town Crier*, 1965a), a subsidy pegged at between K12-K24 per house per annum in 1974-75 was paid to the local authority by the government, in addition to another subsidy of K6 per house per year from liquor undertaking. In 1967, a rental increase was effected and although a new rent policy to account for the full cost of housing was implemented in 1970, its application seems to have been limited to new housing, for example in Kwacha East (Kasongo and Tipple, 1990). Otherwise, no rental increase on old housing was effected between 1967-75 (City of Kitwe Development Plan, Survey of Existing Conditions). A specimen copy of the tenancy card issued to council tenants in 1994 is presented in Figure 8.16. Although KCC was still operating 17 liquor sales outlets in 1991 (KCC, 1991b), limited profits affected the amount transferred to the housing account, resulting in huge deficits, with housing arrears estimated at about K750m in March 1995 (KCC, 1995a).

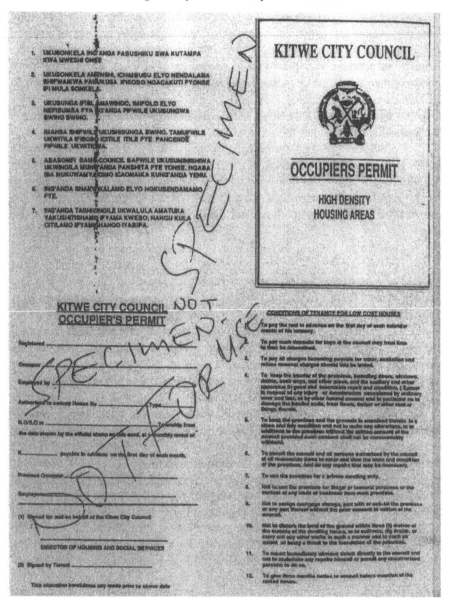

Source: Kitwe City Council.
Figure 8.16   Tenancy card, 1994

Following the change of government in 1991, the new MMD government in 1994 directed all local authorities to sell all their housing stock to sitting tenants. By 1995, 30 per cent of Kitwe city council's housing stock had been sold (KCC,

1995d). While this benefits sitting tenants, disposing of all public housing without building more will not improve the supply of housing, especially if bought for consumption. Unless such housing is released on the market, the supply of housing remains the same. After all, it is the function from consumption to investment which changes when such housing is put on the market, the absolute level of supply still remains unchanged. Although it has improved access to housing for the sitting tenants, the sale of public housing cannot improve the total supply of housing, and indeed may restrict access for other households. For instance, lured by the high rental income, some home-owners might let out their houses and move into cheaper accommodation, increasing demand for low-cost housing and consequently lead to increased values, and also moderating values on former institutional houses being surrendered for re-sale or rental.

In summary, the range of houses developed by the mine company starts from the squalid 20 square metre two-roomed houses in Wusakili, with communal toilets and showers, to four- and five-roomed houses complete with all modern sanitary facilities set in generous surroundings. On the other hand, council housing is mostly two- or three-roomed, except for Chimwemwe and Kwacha East, all others have pit-latrines and outside standpipes. Services such as water, roads and stormwater drains in the mine townships are better than the public township, in which stormwater drainages are not maintained, water supply is erratic, and the once-made roads in some sections of the public township are like the unmade ones in informal settlements. As for rents, mine housing has generally been heavily subsidized, while the local authority has been under increasing pressure from the government to charge an economic rent (City of Kitwe Development Plan, Survey of Existing Conditions).

The council has defined densities of 24-37 dwellings per hectare as low-cost areas, 10-24 as medium-cost and less than ten as high-cost (City of Kitwe Development Plan, Survey of Existing Conditions). Except for Buchi and Chimwemwe, all council and private low-cost housing conforms to these standards (Table 8.4).

Table 8.4   Average nett housing density in Nkana-Kitwe

Classification	Council (existing)	Mines (existing)	FoNDP
Low-cost	31	40	15
Medium-cost	16	27	12
High-cost	6	5	11

Notes: Density in dwellings per hectare. FoNDP = Fourth National Development Plan.
Sources: Compiled by the author from City of Kitwe Development Plan, Survey of Existing Conditions; GRZ, 1989, Table XVDI.5.

Mine low-cost housing on the other hand is generally higher density for various reasons. There was the pragmatic reason to house all its employees within walking distance of the work place, and water-borne sanitation enabled them to increase the density of housing. A similar sense of economy seems to have determined the planning of the mining company's medium-cost housing. The historically European high-cost housing was developed at a density lower than similar housing areas in the public township (except for Nkana East which, although being in the public township, was developed by the mining company to house its own employees). With the council unable to afford the cost of servicing more land, the low densities of Buchi low-cost housing (which, in 1974-75 was earmarked for re-development beginning 1992: City of Kitwe Development Plan, Survey of Existing Conditions) and high-cost areas hold promise for infilling, a possibility examined later. While increasing the density of high-cost housing, the Fourth National Development Plan's (FoNDP) (GRZ, 1989) recommendations proposed lower densities for low- and medium-cost houses (Table 8.4).

Apart from mine and local authority housing, other companies and individuals made their own housing arrangements. Both the Employment of Natives Ordinance of 1929 and the Urban African Housing Ordinance of 1948 placed an obligation on the employer to provide housing for its employees and, in the case of the 1948 ordinance, to the housing of employees' dependants as well. The next section examines this other form of private corporate or individual housing initiatives.

Other Private Housing

The two basic forms of other private housing in Kitwe are company (apart from the mines) and private individual. Because of the prevalence of building contractors in company housing (for example, Amlew, Roberts and MacKenzie), it is also sometimes referred to as contractor housing (Figure 8.17).

These types of dwellings were largely temporary (for the duration of the building contract), and where they have survived, have taken on the character of informal settlements. Other contractor housing in Kitwe was provided in the Kampembwa Industrialists Compound (Figure 8.18).

Rhodesia (now Zambia) Railways established a barrack type single quarters compound near the railway station (Figure 8.19) and other houses spread throughout the town.

Source: City of Kitwe Development Plan, Survey of Existing Conditions, p.37.

Figure 8.17   Robert's compound, Kitwe

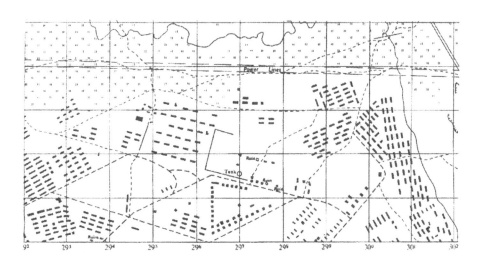

Source: Zambia Survey Department, 1:5,000 Topo-Cadastral Series Map, Sheet No. 2886.

Figure 8.18   Layout of Kampembwa Industrialist Compound, Kitwe

Source: Author, 1997.

Figure 8.19   Zambia Railways barrack-type housing, Kitwe

Another form of private housing is that provided by individuals on land leased from the local authority or private landlords. Before Kitwe Location was established as an African compound in 1942, a suggestion had been made by government that land be alienated and allocated to Africans who would be allowed to develop their own self-help housing. The mining company strongly opposed the proposal, arguing:

> An uncontrolled native village in the near vicinity of a large settlement like Nkana is a menace to the health and safety of the community and is a practice that, in our opinion, cannot be too strongly condemned. Nkana Mine has had bitter experience of the evils of such settlements, having already had to take steps to stamp out certain unauthorized native villages of this nature which grew up in its vicinity (ZNA: SEC 1/1535, Vol. II: Memorandum dealing with the future trading and native housing arrangements at Nkana, p.8).

Although private locations were common in Lusaka and Ndola, Tipple (1978) observes that the only private location (which had disappeared by 1978) on the Copperbelt was Kitwe's Davidson, which had 30 dwelling units.

Nearly all surveyed plots in the high-cost rateable area of Kitwe were developed either by private companies or European individuals and housed in the region of 1,200 people compared to 6,600 in the temporary contractor housing and

41,000 in the local authority's low-cost areas (Retrospect, 1962; Municipality of Kitwe, 1962).

The most significant other private housing was the post-independence site-and-service scheme. In a pragmatic shift from the tied-housing policies of the pre-independence era, the Zambian government sought to promote home-ownership for the many Zambians who had depended on local authority or other institutional housing (for which there was less money to provide).

Following Circular 59/66 on aided self-help housing (Chapter 4), Kitwe started site-and-service schemes in Zambia, Bulangililo (Showpiece) being its first scheme in 1967 with 1,400 serviced sites. Circular 29/68 (Chapter 4) introduced the concept of the basic site-and-service scheme in which only a site and shared standpipe were provided, and three schemes were launched:

(i) Luangwa with 1,600 sites in 1996 in 1970;
(ii) Twatasha with 860 sites in 1996 in 1971; and
(iii) Kawama in 1980 (1,450 serviced sites in 1996 out of a planned 4,500).

A summary of the estimated distribution of the formal housing stock in 1996 by ownership is given in Table 8.5.

Table 8.5 (last column) shows that, while individuals owned a meagre 14 per cent of all formal housing units in 1996, the rest was owned by institutions, public and private, the biggest two being Nkana Division (40 per cent) and the local authority (41 per cent). There are, of course, historical reasons for this state of affairs. Two regulations, the Employment of Natives Ordinance of 1929 (later called the Employment Act) and the Urban African Housing Ordinance of 1948 obliged employers to provide housing for employees only (1929 Ordinance) and their dependants (1948 Ordinance), because poor wages did not allow Africans to adequately house themselves. Wages were so low that in 1944 the Eccles Commission recommended a minimum wage to force employers to pay for married workers' housing (Northern Rhodesia, 1944). Tied housing and low wages hindered individual housing initiatives. Although most individual houses are in low-cost areas, not all of them are owned by low-income households, as will be shown later.

The above section has identified the main agencies, form and level of formal housing supply in Nkana-Kitwe, and the changing form of housing structures have been shown. The next section compares local authority housing supply and demand.

Table 8.5   Estimate of percentage distribution of formal housing stock by ownership in Nkana-Kitwe, 1996

Owner	Low-cost	Medium-cost	High-cost	As % of all units
Mine	31	85	51	40
Local authority	54	4	3	41
Central government	0	1	6	1
Other companies	0	1	13	2
Parastatal	1	3	6	2
Individual	14	6	19	14
Others	0	0	2	0
Total	100	100	100	100
As % of all housing units	76	10	14	

Source: Compiled by author based on sample data from the 1996 Kitwe valuation roll.

Supply of and Accessibility to Land and Housing

Level of Formal Housing Supply and Demand

Housing need refers to the deficiency in dwelling units which should be provided to bring housing to some socially acceptable standard, while housing demand is the expressed desire for housing (UN, 1967). Because of the heterogeneous nature of society, changing social values and the political nature of the housing issue, the definition of housing need is often problematic. Demand, on the other hand, can easily be determined by observing trends in the housing market and council waiting lists - the latter normally acting as an index for demand in the lower sector of the housing market. On their relative merits, Sanyal *et al.* (cited in Schlyter, 1984, p.55) note 'housing should be based on effective demand not on politically defined needs', while the City of Kitwe Development Plan explains:

Estimates of effective demand are important in formulating a housing policy in that the extent to which households in need of housing can pay for improved housing, will determine the extent to which the housing programme can be expected to be financially self-supporting (City of Kitwe Development Plan, Survey of Existing Conditions, p.63).

The extent to which housing need exceeds demand reflects the population needing housing but perhaps unable to afford the cost and hence requiring some form of housing subsidy. Effective demand therefore is important when considering the level of policy sustainability from a financial and political standpoint. Public policy which excludes a sizeable portion of the population is likely to be unpopular.

Table 8.6 shows the level of local authority housing against the demand. The housing stock figures between 1966-70 refers to all housing units owned by the local authority, and between 1974-95 to low-cost housing only. It is not expected that there would be a significant difference between the two as to affect this analysis, because the local authority owns a meagre three per cent of Kitwe's high-cost housing compared to 54 per cent of the low-cost (Table 8.5). Housing demand refers to the number of applicants on the local authority's waiting list.

Table 8.6 Kitwe City Council housing stock and demand, 1966-95

Year	Housing stock	Demand
1966	10,434	8,442
1967	11,168	12,965
1968	11,286	17,588
1969	11,718	19,277
1970	13,726	21,479
1974	13,324	26,000
1981	13,325	11,072
1983	13,289	13,300
1985	13,289	20,000
1991	13,275	19,500
1995	13,166	23,500

Sources: C.S.O., 1971; NHA, 1974; various local authority reports.

In 1966, two years after independence, demand for formal housing was estimated at about 8,400, rising sharply within two years to 17,600. The high peak of 26,000 in 1974 (an increase of three times the 1966 level) can be partly explained by the boundary adjustment of 1970 in which outlying districts of Chambeshi, Kalulushi and Chibuluma were integrated into the city of Kitwe. By 1980, three site-and-service schemes had been implemented and the above outlying districts were excised from Kitwe, hence the drop in demand to 11,072 in 1981 (only 31 per cent higher than the demand in 1966). With the council's rental housing development discontinued in 1972, effective demand rose steadily from 1981 to 1983, more steeply from 1983 to 1985, stabilizing then tending upwards after 1991. By 1995, demand was almost three times the 1966 level.

On the other hand, the number of local authority houses increased from 10,434 in 1966 to a high of 13,726 in 1970, representing a 32 per cent growth over the number of houses in 1966. While the general demand trend has been upward, the housing stock has remained almost static since 1970.

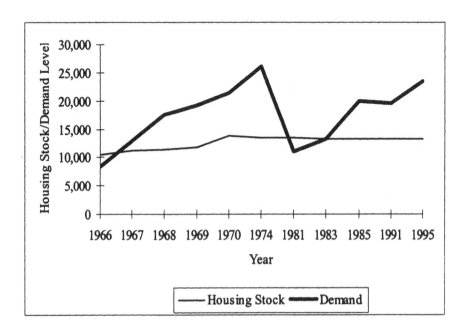

Sources: C.S.O, 1971; NHA, 1974; various local authority reports.

Figure 8.20 Kitwe City Council housing stock and demand, 1966-95

Thus the supply of local authority housing has not matched demand. Although by 1995 demand was about 180 per cent over the 1966 level, the housing stock has only risen by 26 per cent. Largely due to this failure of the public housing programme to meet demand, the concept of home-ownership on the self-help basis of site-and-service schemes was conceived. As this aimed to address the rising demand for housing, it is appropriate to evaluate the extent to which sites were allocated to all social groups, with access to land and housing finance the most commonly cited constraints to housing (UNCHS (Habitat)), 1974; Angel *et al.*, 1983; Payne, 1989; Gilbert, 1992).

Supply of and Accessibility to Land for Housing: Cost Aspects

Availability of and access to land and finance are claimed as the major constraints to housing provision. The two are affected by a number of factors: land tenure, the cost of land and housing as a proportion of household income and aspects of land administration related to land subdivision, land allocation and development control. Chapter 4 gives an account of national aspects of land tenure, policy and administration. A detailed discussion of the cost aspects of land and housing follows, using, as far as possible, local empirical data on Kitwe.

Costs in relation to service and construction standards: Kironde (1992) and Okpala (1987) censure the use of foreign standards in Africa's urbanization process. They argue that these need to be re-defined to reflect an understanding of local needs and resources. Before proceeding to look at how service and construction standards have affected access to land and housing in Kitwe, it is important to contextualize the issue. Kitwe was developed as a European town (the African presence was simply tolerated), the European sector of the city being laid out at a generous density away from African locations. Clean piped water was provided to both the European and African sectors, albeit at different standards. The indigenous African was exposed to foreign municipal and building standards during this time. It is therefore not surprising that the first post-independence site-and-service scheme in Kitwe called Bulangililo was designed to a high standard, attracting a high-income population at the expense of the poor (as will be shown).

High standards can be seen in other site-and-service schemes. In Twatasha for example, less than a quarter of houses had been completed 14 years after the scheme had began, and all completed houses were made of concrete blocks (Todd and Sinjwala, 1985). Since the launch of Kawama in 1980, a scheme subsequently entered as a show case of good practice in the UNCHS (Habitat) competition (Copperbelt University Shelter Study Team, 1995), only 1,459 plots out of a total planned of 4,500 were serviced and allocated by 1995 (KCC, 1995d), with high standards and careful selection criteria at the expense of not only the poor who cannot afford the high standards, but also those that can but have not been able to get an allocation because the council has not serviced the plots.

In spite of the fact that standards in Luangwa were relaxed and people allowed to use Kimberly brick for the super-structure, some affluent households used concrete blocks, thus putting pressure on poor households who also wanted to use better materials but could not afford to. In 1996, 26 years after the scheme had been launched, 55 per cent of the plots in Luangwa were still undeveloped. People's standards are influenced by what they see of other households or settlements, so the argument about basic standards should not be patronizingly used to stifle people's aspirations for better services if they can afford them. While providing or facilitating basic housing and services for the poor, better standard housing helps the poor aspire to a better life. As we have seen in the case of Luangwa, the bias for better standards is not limited to official thinking. One of the complaints received from residents in Kitwe's site-and-service schemes is that the plots are 'squeezed together and have no space for gardening and parking vehicles' (KDC, 1983), suggesting that the rich, accustomed to the luxury of space, are acquiring plots in site-and-service schemes.

The official commitment to high standards in Kitwe has raised the costs of services and construction. In almost 20 years after proposals for upgrading (but actually a complete re-development), two informal settlements, Kamatipa and Ipusukilo, still remain undone (Kasongo and Tipple, 1990).

Cost in relation to household income and local authority revenue: The traditional approach to the public supply of urban land for development in Zambia has been that of plan - survey - service -allocate and develop. As Rakodi (1991) observes, the public supply of land cannot be discussed without reference to services as these affect availability, cost and affordability. Even when services are not provided before allocation (as has increasingly become the case), the local authority still assumes that it will provide these at a later date and includes them in the cost of land. In the informal settlement, however, access to land is influenced by urban and traditional attitudes. An initial claim is established if not contested, or permission is sought from an identified owner or authority - normally a political figure in post-colonial Zambia (Mutale, 1993). Services are provided later through the community's own efforts, or by the local authority through lobbying. Notice the reversal of activities between the formal and informal approach. The advantage with the informal approach is that the initial expense goes towards the construction of shelter, as opposed to the formal approach which imposes high upfront costs.

Kitwe was the largest benefactor of housing funds dispensed by government through the African Housing Board in 1965 (GRZ, 1965). In 1967, the first site-and-service scheme in Kitwe (Bulangililo) was launched under the FiNDP. As its 'showpiece' status implied, the quality of services was high, reflecting standards promulgated in Circular 17/65 (Chapter 4). Plots averaging 380 square metres in size were fully serviced with an internal water supply and water-borne sanitation, at a cost of K400 (£200) per plot. It soon became apparent that this level of servicing was unaffordable by the lowly paid workers, and the concept of basic site-and-service was adopted in 1968 as described in Circular 29/68 (Chapter 4). By limiting services to a communal water supply, with plot sizes reduced to about 300 square metres, costs were cut from K400 to K140 with a monthly rental of K1.75, and such basic site-and-service schemes were launched at Luangwa (1970), Twatasha (1971) and Kawama (1980). To determine the extent to which low-cost plots are accessible to the poor, four measures are used:

(i)   examine the values of houses in these areas and relate these to official estimates of low-cost housing (Table 8.7);
(ii)   examine the ability of plot owners to develop (Table 8.8);
(iii) examine the distribution of housing across the social spectrum; and
(iv)   compare the ratio of income and house costs for all income groups.

Table 8.7   Official and observed low-cost values in Kitwe, 1989

Locality	Official estimate (K)	Observed average value (K)
Bulangililo	17,600–42,800	85,000
Luangwa	as above	53,460
Twatasha	as above	76,000
Ndeke H	as above	117,690
Ndeke G	as above	86,000

Sources: Official estimates from GRZ (1989) and observed values from Kitwe 1989 valuation roll sample.

From Table 8.7, the observed value of housing in all areas shown is higher than the official estimate for the same type of housing, within the officially estimated medium-cost band of K42,800-K167,000 (GRZ, 1989). It is possible that the official estimates are not realistic. If the official estimates are realistic and yet observed house values compare with official medium-cost estimates, then it seems that these schemes have not benefited the intended target group. While initial plot allocations might have been to the poor, plots may have subsequently been sold to others with a higher income. Tipple (1978), referring to Luangwa and Twatasha, observed that most original owners have left the schemes. If most original owners sold out, and yet an examination of change of ownership between 1989 and 1996 (Chapter 6) reveals few such changes, these ownership changes seem to have taken place shortly after the inauguration of these schemes, in 1970-71. The relatively high level of services in Bulangililo makes it probable that these plots originally appealed to the high- and middle-income groups, who proceeded to develop housing to a higher standard than the plots were allocated for. Rakodi (1991) observes similarly for Lusaka. It might be argued that the high values simply reflect the stringent building control imposed on the owners, but this only strengthens the argument that the owners not only met the minimum standards, but developed to an even higher standard. The high values in Bulangililo were also observed by the council's own survey of low-cost housing (KCC, 1976).

While all low-cost home ownership schemes referred to in Table 8.7 show a higher observed value than the official low-cost estimate, the small number of undeveloped plots in Bulangililo compared to Luangwa suggests that its superior services and proximity to the CBD and other social services attracted a high-income population, either as original allottees or subsequent buyers with the means to develop. Luangwa, on the other hand, is a basic site-and-service scheme on the periphery of the city and although initiated soon after Bulangililo (three years later), 55 per cent of plots in Luangwa still remain undeveloped (Table 8.8).

Table 8.7 shows that although higher than the official estimate, housing in Luangwa has lower values than other schemes, probably dampened by the settlement's remoteness and poor services. The inferior housing and failure to develop 55 per cent of the plots in Luangwa reflect the low-income status of plot-owners in Luangwa, where no selection based on financial capability was done. The entire population from the nearby informal settlement of Chibili was re-settled in Luangwa (Tipple, 1978; Todd and Sinjwala, 1985). Councillors and local authority technical staff were not enthusiastic about basic site-and-service schemes (Kasongo and Tipple, 1990), and subsequent schemes in Twatasha and Kawama, although classified as basic, were serviced to a higher standard than Luangwa, an attitude common to most local authorities, as Rakodi (1991) observes for Lusaka. When Twatasha and Kawama were launched, the selection procedure sought to allocate land to those with the financial means to develop, either formally or self-employed with savings of at least K500, with preference to council tenants and those in informal settlements.

Table 8.8   Development of low-cost home-ownership schemes, 1989 and 1996

Scheme	Developed	Undeveloped	Undeveloped as percentage of total
1989			
Bulangililo	688	12	2
Luangwa	344	451	57
Ndeke G	197	67	25
Ndeke H	208	99	59
Twatasha	407	2	5
1996			
Bulangililo	691	6	1
Luangwa	364	439	55
Ndeke G	198	66	25
Ndeke H	254	257	50
Twatasha	415	21	5

Source: Compiled by author from 1989 and 1996 Kitwe valuation rolls.

Another determinant of accessibility to formal land and housing is to examine the distribution of ownership. A summary of home ownership levels in Table 8.5 shows that individuals owned only 14 per cent of the housing stock in Nkana-Kitwe in 1996. The provision of tied-housing and low wages has hindered formal individual housing initiatives, especially among indigenous Zambians.

So far, proxy measures have been used to determine accessibility to housing land by the poor. An attempt is now made to determine directly the cost of housing land and house development costs in relation to household income and local authority finance. It could be argued that officially land has had no value, but it was undeveloped land which, between 1975 and 1995 was considered to have no value. To the extent that costs of services were included in the cost of residential plots as a service charge or as a monthly rental, in the case of site-and-service schemes, this cost component can be valued. For high-cost plots, service charges would normally be paid upfront and land costs in site-and-service schemes were factored into monthly rental repayments. Assessment of applicants would include financial capacity to pay these costs and build the house.

Table 8.9   Land values as a percentage of improvement value

Housing classification	1954	1963	1970	1989	Average in 1989
Low-cost	--	--	--	17.3-26.0 (10.0)	22
Medium-cost	--	13.7	10.0	10.2-29.9 (20.0)	20
High-cost	17.2	16.6	14.9	23.7 (20.0)	22

Note: While 1954, 1963 and 1970 values are based on sample data for Kitwe, 1989 values in brackets are Kitwe estimates from Hammar (1990), who uses international experience. The other 1989 values are national rather than Kitwe estimates from GRZ (1989).
Source: Various.

The exemption of low-cost housing from rates (Northern Rhodesia, 1957a) until 1989 precludes a historical assessment of land costs, as low-cost land and housing does not appear in valuation rolls prepared before 1989. Land as a percentage of house values in 1989 showed no significant difference between the residential sub-groups, being about 20 per cent of house value for all housing areas (Table 8.9 last column). Whereas high- and medium-cost land value components have risen from 17 per cent and 14 per cent respectively to 20 per cent, there is no historical data to show the same for low-cost housing. If the same trend can be established for low-cost areas, it would show a convergence in the residential land market which would prove the point that the rich are buying into poor areas. The higher the proportion of land value, the more poor people get pushed out of the low-cost land market. A satirical cartoonist (Figure 8.21) in a Zambian independent newspaper captures this scenario, albeit in a different sort of market.

Source: The Weekly Post, 13-19 March 1992.

Figure 8.21   What are market forces?

At 20 per cent of house development cost, is land expensive in Kitwe? Table 8.10 translates this percentage into monetary values and relates the same to available housing expenditure in 1991.

The government survey (GRZ, 1991b) gives detailed information on incomes and expenditure and, despite falsification (KCC, 1976; Okpala, 1987), still provides the only (and, at the time of writing, complete and up-to-date) source of empirical data to compare housing costs with income. With 17 per cent of monthly income available for housing in 1991, a low-cost plot was worth ten years' savings. If a family aspired for a medium-cost plot and saved 23 per cent of its income per month, it would need to save for 18 years before it could afford the plot; at 17 per cent of monthly income, a high-income household would need to save for 55 years! Is it any wonder that there has been a down-market movement by the rich? In 1978, Kitwe's local authority proposed to upgrade six informal settlements at a cost of K103 (£40) per plot, at a time when two thirds of the national urban population earned less than K100 per month.

Table 8.10 Land and improvement costs in relation to available housing expenditure in Kitwe, 1991

Housing classification	Low-cost	Medium-cost	High-cost
Average monthly income (K)	9,579	11,901	12,786
Housing expenditure as percentage of average monthly income	17	23	17
Monthly housing expenditure (K)	1,628	2,737	2,174
Estimated total housing cost (K)	957,964	2,940,000	7,160,000
Estimated land cost at 20% of total housing cost (K)	191,593	588,000	1,432,000
Estimated improvement cost (K)	766,371	2,352,000	5,728,000

Sources: Income and expenditure from GRZ (1991b), housing costs estimated by the author from 1989 and 1996 valuation rolls.

These projections may be misleading for a variety of reasons. Housing expenditure in the government survey is defined as including all rental, water and fuel costs, with no reference to land. Because the family needs to rent shelter and pay for water and fuel, the amount set aside for land would have to be in addition to the housing expenditure, and would represent an extra 17-23 per cent of household income per month. In terms of annual household incomes, total housing costs work out to eight, 20 and five times total annual household income for low-cost, medium-cost and high-cost respectively.

The problems of using such notional percentages as a basis for policy are compounded by instability in most developing world economies. Although in 1991 it was reported that household expenditure on housing averaged 22 per cent, the situation was radically different in 1993 and 1994 (Table 8.11).

As the real earning power of incomes declined, households had to trade-off housing expenditure with other needs. A 1993/94 household budget survey shows that real urban incomes declined by 47.8 per cent between 1974 and 1994 (GRZ, 1995c), and household expenditure items in GRZ (1991b) no longer included savings. The total aggregate of household expenditure comprises food, housing, clothing, transport, remittances, education and medical care. Personal experience shows that it is impossible to expect the majority of households to put aside on a regular basis 17-23 per cent of their income towards the purchase of a residential plot.

Table 8.11 National urban percentage household expenditure on housing by socio-economic group, 1991-94

Housing Classification	1991	1993	1994
Low-cost	17	12	8
Medium-cost	23	10	--
High-cost	17	10	13

Sources: GRZ, 1991b, 1994 and 1995b.

Because the use of reported household income and expenditure as a measure of access to land and housing is fraught with problems (Okpala, 1987), the proxy measures above give reasonably hard evidence on the inability of poor households to pay for land and housing. Given that individual household priorities and circumstances differ, who is to say that land costing under a certain theoretical value is cheap? What is cheap will largely be a matter for individual households to decide, as they weigh the relative importance of various needs competing for limited resources.

Cost constraints have also hindered the local authority in servicing land for development at Kawama and Riverside Extension phase 4C. In the latter, at the lower end of the high-cost housing market and classified in some government publications (GVD Quarterly Bulletins) as medium-cost, the cost of servicing 324 plots was estimated as K183m in April 1992 (approximately K565,000 per plot and compares with estimated land costs in 1991 for medium-cost plots in Table 8.10). By September 1995, the local authority's Land Development Fund (Chapter 4) account was worth a paltry K6m, when the 1992 estimate for servicing the 324 plots had almost trebled (KCC, 1995d). Requests from plot-holders in Riverside Extension for plots in other serviced areas could not be granted because the council did not have vacant serviced land. In 1996, 310 plots in Riverside Extension still awaited development (Sample Survey, 1996 valuation roll), and only 1,450 of the planned 4,500 sites in Kawama had been serviced, 16 years after the scheme was launched in 1980.

Until independence, the predominant form of housing was that provided by institutions, notably the mining company and the local authority. Although the latter's housing stock grew from about 1,140 in 1953 to 10,000 in 1964, most of this was either tied to a job or regular income, discriminating against others for whom, as Tipple (1981, p.74) observes '...in British society council housing provides an important source of shelter'.

Public housing development could not match the demand and had, in any case, ceased by 1972. Increasing financial constraints led to the introduction in 1967 of self-help initiatives in the form of site-and-service schemes, the local authority providing serviced sites sometimes with building material loans, and owners expected to build their own houses. A detailed examination of cost aspects has revealed that, in spite of the good intentions of these schemes to address the housing needs of the poor, cost constraints restricted accessibility by the poor. Even in instances where the poor have managed to get a plot, these were subsequently re-sold or remained undeveloped. Notwithstanding progress in the level and form of housing, especially mine housing, formal housing initiatives have had limited success in meeting demand. The following section traces the origins and growth of informal housing, and compares levels of housing with the public low-cost housing sector.

Emergence and Progress of Informal Housing

Rigorous migration and settlement control meant that at independence there were no informal settlements within the built urban area of Kitwe (Tipple, 1976a). A government report (Northern Rhodesia, 1944) estimated that there were about 200 married men and 200 single men living in unauthorized settlements around Kitwe in 1944. The first known surviving informal settlement is Mufuchani, east of the Kafue River, established in 1952 (Figure 5.10). In 1969, only Kampembwa (Figure 8.18) and Chimwemwe North showed a concentration of informal settlements, the former being removed by 1976 to allow for industrial development (Todd and Sinjwala, 1985). While informal settlements were almost non-existent within the built-up area of Kitwe, most housing, as has already been discussed, was owned not by individuals but by institutions. Table 8.12 illustrates the growth of population in informal housing between 1969-98 in relation to the entire population of Nkana-Kitwe.

Table 8.12 shows that Nkana-Kitwe's informal housing population grew from nothing at independence to 60,000 in 1969, accounting for 30 per cent of the city's population. With the establishment of two additional site-and-service schemes at Luangwa in 1970 and Twatasha in 1971, informal housing population fell to 14 per cent of the total population, but grew to a high of 31 per cent in 1982. While indications are that the informal population has been on the decline since 1982, a drop which could be attributed to the net external migration recorded for Nkana-Kitwe in the decade 1980-90 (GRZ, 1995a), it still constituted about 19 per cent of Nkana-Kitwe's population in 1991 and rose to 26 per cent in 1998.

With limited public resources, the local authority responded to the growth of informal housing with site-and-service schemes: Bulangililo in 1967, followed by Luangwa, Twatasha and Kawama.

Table 8.12 Nkana-Kitwe's informal housing population, 1969-98

Year	Number of settlements	Informal housing population	Total Nkana-Kitwe population	Informal housing population as percentage of total population
1969	--	60,000	199,800	30
1973	40	35,000	249,000	14
1974	--	47,400	336,820	14
1978	63	55,000	260,000	21
1981	--	88,270	314,790	28
1982	28	108,200	350,000	31
1991	22	62,930	338,210	19
1998	22	88,462	343,100	26

Sources: NHA, 1974; Van den Berg, 1974; KDC, 1981; Kasongo and Tipple, 1990; Kitwe Water Supply Project, 1991; Hansungule *et al.*, 1998.

While Luangwa and Twatasha were being launched, Kitwe City Council's Town Clerk presented a radical paper to the Town Clerks' Society of Zambia (Anon, 1971) opposing the idea of basic site-and-service schemes, claiming that people were migrating to urban areas lured by almost free services. Stop the provision of subsidized basic services, and this will stem the tide of migrants. While arguing that even the basic services provided in urban areas were an improvement on the poverty of rural areas, he proposed that informal settlements be moved away from present developed municipal areas, sited outside a radius envisaged as potential development zone for the next ten years and provided with a treated well only. The denial of basic services would not stop the migrants because of other pull factors, for example real or perceived job opportunities. The Town Clerk's thinking is underlain by material value assumptions, and his reference to the irresponsible character of Zambian society fails to recognize underlying colonial political views and economic factors such as the migratory view of African labour, wage disparities, and paternalism which have influenced cost recovery in service provision and home-ownership. Tipple (1976b) suggests that people are happy to live in unserviced areas at no apparent cost rather than submit to the control of and pay for services in formal housing. Once these people are in town, it is immoral and politically unacceptable to let them succumb to disease and death by failing to provide them with basic services. The potential for negative external effects warrants some form of provision, however it may be financed and charged for. There is value in the Town Clerk's global view of the squatter problem. Unless attempts are made to address the root cause of the problem (which he identifies as rural-urban migration), we will forever be dealing with the consequences of a problem whose roots lie somewhere else, a waste of resources which could be channelled to addressing the fundamental causes. The issue however is, to what extent is the squatter problem a migration issue? If it is not, how successful will attempts be to repatriate to rural areas generations whose only life experience is urban?

The resettlement of informal households into site-and-service schemes has had its problems. Attempts to relocate Nkandabwe and Saint Anthony (informal settlements threatened with mining subsidence) to Kawama have failed (Mulwanda and Mutale, 1994; ZNBC, 1994). Although most households in Nkandabwe relocated, a few still remain (*Times of Zambia*, 6 August 1997, 'Opinion'). Tipple (1976b) commented:

> ...[the] squatter resettlement idea has not been successful. It was based on a false assumption that squatters would want to move to a serviced area and pay for the services provided rather than live in an unserviced area for no apparent cost...the control and cost elements involved in a site-and-service scheme seem to have proved sufficient disincentive to the majority of potential movers (Tipple, 1976b, p.168).

The other cost element involved in the failure of squatter resettlement policies is that of transport. Most informal settlements have located near employment sources, and residents are not willing to move to distant places. Resettlement was also frustrated by selfish politicians who feared to lose their constituencies (Todd and Sinjwala, 1985; Mulwanda and Mutale, 1994; *Times of Zambia*, 6 August 1997, 'Opinion').

Because of the difficulties in persuading people to resettle in serviced areas, and the realization of the economic value of existing informal housing, the SNDP, while putting an end to public rental housing and emphasizing home-ownership on serviced sites, also provided for the upgrading of informal housing through the Housing (Statutory and Improvement Areas) Act of 1974 (Chapter 4). A policy statement on informal settlements in Kitwe proposed to group the 45 odd settlements into six main settlements for upgrading (Kamatipa, Mulenga, Luangwa II, Kalibu, Ipusukilo and Itimpi) (KCC, 1973b) still remains a statement of intent. Apart from a few feeble efforts, there has been no meaningful attempt at upgrading in Kitwe. In a bid to forestall a possible cholera outbreak, emergency water supplies were extended to ten informal settlements in 1975 (Kasongo and Tipple, 1990). Community efforts are now being used through approaches such as Project

Urban Self Help (PUSH) to upgrade four of the six settlements. Itimpi, Ipusukilo and Race Course have already benefited from the PUSH programme (Hansungule *et al.*, 1998, Kitwe Evidence).

While the informal housing sector grew from almost nothing at independence to a situation in 1981 of 100 informal dwellings for every 97 local authority low-cost units (Table 8.13), by 1998, of every 100 informal dwellings, the local authority provided another 89.

Table 8.13 Local authority low-cost and informal housing units, 1974-98

Year	Local authority (formal)	Informal	Ratio of informal : formal
1974	13,324	9,800	1 : 1.36
1981	13,325	13,792	1 : 0.97
1983	13,289	15,072	1 : 0.88
1985	13,289	--	--
1991	13,275	10,236	1 : 1.30
1998	13,275	14,907	1 : 0.89

Sources: Various KCC Annual Reports, NHA, 1974; Kitwe Water Supply Project, 1991; Hansungule *et al.*, 1998.

Definition of the Housing Problem in Nkana-Kitwe

The discussion so far has shown that the deficit in local authority housing is not confined to the post-colonial era, but has its roots in the period before independence. This deficiency in formal housing has been met through informal housing provisions and an expanding rental market (Kasongo and Tipple, 1990). The reality and extent of housing need was shown dramatically in 1974 when a contractor's settlement called 'Charlie West' serviced with three water taps grew from 19 dwellings to 1,800 in just four months, mostly with squatters from an adjacent settlement and sub-tenants from a nearby council area (Tipple, 1976a). The evidence of the growing waiting list is treated as demand rather than need. Before an attempt is made to define the current housing problem in Nkana-Kitwe, and because of the link between population and housing, population trend is analysed.

Table 5.2 (Chapter 5) showed that the population of Nkana-Kitwe rose until 1975 but, except for 1987 (see note b under Table 5.2), not significantly since, and sometimes has dipped below the 1975 level. A number of reasons may have contributed to this situation, among them migration, fertility and mortality. A study on internal migration and urbanization estimated that between 1980-90 Nkana-Kitwe experienced a net out-migration of approximately 54,000 people (GRZ,

1995a), while population increase between 1967-75 was 20 per cent over that predicted (Tipple, 1976a). Data on fertility and mortality are not readily available but with the AIDS epidemic claiming the most productive of the population, low fertility and high mortality may have significantly contributed to a near-static population. The short-term population trend of 1974-90 has been extrapolated with a linear regression (Oppenheim, 1980) model computer programme by Ottensmann (1985). It has been decided to use this rather simple approach because, not only is it widely used in planning, it is also less demanding on data compared to other more robust but data intensive models. Because conditions defining the 1974-90 trend may not be long-term, it is necessary to project for current needs only, and so only the 1998 population has been projected. The following discussion indicates in broad terms the size and nature of the housing problem in Nkana-Kitwe.

Housing need (defined earlier) refers to the deficiency in dwelling units which should be provided to bring housing to some socially acceptable standard. Given the heterogeneous nature of the urban population and the difficulty of establishing a common acceptable housing standard, a socially acceptable standard of housing in this research is taken from Zambia's Draft Housing Policy, representing the official if not the public view.

> Housing therefore is a priority to government for it ensures that its citizens have decent shelter which is secure and promotes health and well being of communities... culminated into the formation of squatter compounds and slums... (GRZ, 1995b, p.1).

> Decent housing cannot be achieved without the simultaneous development of infrastructure services such as water supply, sanitation, roads, storm water drainage, electricity and others (GRZ, 1995b, p.7).

The above extracts make it clear that informal housing, with its lack of *de jure* security and limited social and physical infrastructure services, is not socially acceptable by official Zambian standards. In quantifying the size of the problem, socially acceptable housing will be limited to formal dwellings only.

In 1991, the total number of formal housing units in Nkana-Kitwe was estimated at 41,530 (Kitwe Water Supply Project, 1991). From information made available to this author from the council's building inspectorate, between 1991-96, 1,893 new residential planning applications were approved. Assuming that only one-quarter (473) of these have been developed, a total formal housing estimate in 1998 is 42,000 (Table 8.14). With a projected population of 343,100 and a household size of 6.4 (GRZ, 1994), the current housing need in Nkana-Kitwe is estimated at 11,600 (Table 8.14), a deficiency in formal housing being partly met through informal dwellings, which numbered about 15,000 in 1998 (Table 8.13).

Table 8.14 Housing need in Nkana-Kitwe, 1998

Year	Population estimate	Estimated housing units needed	Existing formal housing units	Housing Need
1998	343,100	53,600	42,000	11,600

Source: Author's calculations.

This 1998 estimated need is nearly 50 per cent lower than the demand for housing in 1995 (Table 8.6) only three years earlier. Given that the local authority has not built any new housing since the mid-1970s, there are several possible reasons why the estimated need is lower than demand:

(i)   An over-estimate in household size will result in an underestimate of housing need. By comparison, NHA (1974) quotes a figure of 5.5; KCC (1976) estimates 5.3; Knauder (1982) gives 6.3; and lastly Kitwe Water Supply Project (1991) uses a figure of 6.6. A household size of 6.4 used in estimating need compares reasonably with the 1991 estimate above.

(ii) An over-estimate in existing dwelling units will result in an under-estimate of housing need. NHA (1974) estimated existing formal housing stock in Nkana-Kitwe only (excluding outlying districts) as 37,000; the City of Kitwe Development Plan (Table 3.5) quotes a figure of 33,156. Although the NHA and the Development Plan estimates were made at about the same time, the difference of 3,840 might be attributed to the latter's omission of servant's quarters, which the former includes as separate units in their own right, estimating these to be 2,400, although Tipple (1976a) puts the number of servant's quarters at 3,500. Tipple (1976b) cites 31,500 low-cost housing units in 1974. In 1991, it was estimated that of 51,768 total houses in Nkana-Kitwe, 41,532 were in the formal sector, and 10,236 in the informal (Kitwe Water Supply Project, 1991). A more recent study (Hansungule *et al.*, 1998) puts the number of houses in the formal sector at 40,000.

(iii) An under-estimate in population would reduce the quantity of housing need. The projection of current (1998) population from the 1974-90 trend assumes that the conditions establishing the trend are continuing.

(iv) Demand could be inflated by the retention on the waiting list of applicants who have been formally housed privately or have moved to other towns.

(v)   Demand may also be inflated by applicants housed in employer housing who have applied for own council housing. It has not been possible in this research to estimate the number of such applicants.

From the above, it can be surmised that, rather than an under-estimate in need, it is demand which is inflated. Although an up-to-date waiting list may reflect effective demand, some applicants were reported in 1981 as having been on the waiting list since 1964 (KDC, 1981). Based on a household size of 6.4 and a projected population of 343,100, it is estimated that there existed a housing need of about 11,600 units in Nkana-Kitwe in 1998.

Practical Suggestions on Improving Access to Land and Housing

Recommendation (i) - Multiplication by division: In discussing financial aspects of housing land supply, reference was made to the local authority's financial inability to service more land, and the savings made by reducing service standards and plot sizes. Densification through the sub-division of existing plots could provide more land in existing serviced areas and share costs, likely to be welcome to the relatively lowly-paid Zambians now living in the once affluent European high-cost areas. The phrase 'multiplication by division' is taken from Bull (1992) proposing the sub-division of existing low-density plots in Lusaka. Official documents (Local Government Act, Public Health Act and TNDP), give the following housing standards:

Minimum plot size (12m x 27m) = 324 square metres
Maximum developable area (33.3 per cent of plot size) = 108 square metres
Minimum size of living room = 10.8 square metres
Minimum size of bedroom = 8.1-8.4 square metres
Veranda = 6 square metres
Kitchen = 6 square metres

The Draft Housing Policy notes that less than ten per cent of the Zambian population can afford the cost of the above minimum urban shelter. The government buildings branch classifies low-cost housing as 40-50 square metres of habitable space. Using this and the above standards this translates to:

1 living room = 10.8 m^2		1 living room = 10.8 m^2
2 bedrooms = 16.8 m^2		3 bedrooms = 25.2 m^2
veranda = 6 m^2	OR	veranda = 6 m^2
kitchen = 6 m^2		kitchen = 6 m^2
Total = 39.6 square metres		Total = 48 square metres

With an average urban household size of six persons, and using the living room as sleeping space at night (a common practice in Zambia), the minimum reasonable shelter is the two bedroom house with an occupancy rate of two persons per room. Because of this multiple occupancy which is considered normal for single-sex children, room sizes lower than 8.4 square metres although economical, might compromise health considerations.

Research evidence on site coverage ratios in Kitwe's site-and-service areas shows that only 16-28 per cent of the plot areas have been built up for plots averaging 320 square metres (Table 8.15).

Table 8.15 Site coverage ratio, plot size and built-up area in select areas of Nkana-Kitwe

Residential area	Site coverage ratio (%)		Plot size (m²)		Built-up area (m²)	
	Maximum allowable	Actual	Minimum allowable	Actual	Maximum allowable*	Actual
Low-density						
Riverside	25.0	13	1,022	1,621	405	214
Parklands	25.0	14	1,022	2,116	529	294
Kitwe Central	25.0	19	1,022	2,782	696	528
Zebra Street	25.0	15	1,022	2,100	525	306
Riverside Extn.	25.0	34	1,022	1,055	264	357
Nkana East	25.0	12	1,022	1,648	412	203
Nkana West	25.0	8	1,022	2,483	621	191
Average		16		1,972	493	299
High-density						
Bulangililo	33.3	16	269	379	126	62
Twatasha	33.3	19	269	317	105	59
Ndeke H	33.3	28	269	300	100	84
Ndeke G	33.3	25	269	300	100	74
Luangwa	33.3	16	269	305	102	48
Average		21		320	107	65

Notes: * = 25 or 33.3 per cent of actual plot size
All high-density areas in Table 8.15, Riverside Extension and Zebra St. were developed after the publication of the Kitwe Development Plan, the rest were built before.

Sources: City of Kitwe dated about 1967; various individual property files from Kitwe City Council's Valuation Department.

Given that high-density site-and-service built-up areas fall below the maximum allowed standard of 33.3 per cent (or 107 square metres on an average plot size of 320 square metres), savings can be made by reducing plot sizes. A reduction to 170 square metres from the current NHA official minimum of 12 metres by 27 metres (or 324 square metres) allows the development of a three-bedroomed house occupying a total area of 48 square metres (28 per cent of the plot), leaving the remainder for vegetable gardening, an important supplement for low-income households (Sanyal, 1981; Chiwele, 1992). With low-income housing now liable to rates on both land and buildings, rates need to be kept manageable by limiting plot areas to the functional minimum, while allowing flexibility in the design and orientation of the house.

On the other hand, research evidence in Nkana-Kitwe shows that except for Riverside Extension, plots in other low-density areas have been built to a lower density than the stipulated 25 per cent site coverage ratio for this type of development (Table 8.15). Of the average plot area of 1,972 square metres in low-density areas, only about 300 square metres has been developed, representing an average actual site coverage ratio of 16 per cent as compared to the allowable 25 per cent - falling short of the maximum allowable on such a plot by an average of 194 square metres (493-299 square metres). Given that 14 per cent of the estimated 42,000 formal housing units are in high-cost and therefore low-density areas, by amending density zoning regulations in low-density areas to allow more dwellings per hectare, and reducing current plot size standards to the functional minimum, potential exists to provide in the region of 6,000 mostly low-cost and a few medium-cost additional serviced plots, having regard to the limits in the capacity of existing services.

Because of the cost constraints in land and housing, the sub-optimal development of existing residential plots, and given the economic benefits for the local authority and residents from reduced service costs and cost sharing, the FoNDP proposals can be criticized. While increasing the density of high-cost areas, they propose a comparatively lower density for low- and medium-cost houses than existing (Table 8.4). Given the prevailing financial stringency and the urgency of supplying housing development land at an affordable rate, should not local authorities promote higher density development? Densification through sub-division would occur through the market since these are private plots.

Recommendation (ii) - Recognizing and supporting the market supply of land and housing but...: The responsibility for identification, subdivision and allocation of formal housing land in Nkana-Kitwe and the rest of urban Zambia normally lies with the local authority. While private individuals may subdivide and sell land through the market, the public and official attitude is that councils provide housing land, and so private subdivisions have not been considered as a viable proposition for providing land for housing, despite evidence in Kitwe on the ability of the market to match the public supply of urban land and property (Chapter 6). Although the market is often labelled as exploitative and hence discriminatory to the poor, public allocation has its flaws: councillors and ward chairmen allocated plots in Kawama contrary to administrative procedures (Todd and Sinjwala, 1985).

A public supply of land resulting from densification in (i) requires local authorities to negotiate and pay individual owners, and then allocate to the public. A public allocation, however, is likely to increase the delivery time and cost of the plots by introducing an intermediary between the owner and potential buyers. Even if councils wanted to use compulsory purchase powers, it is unlikely in the current Zambian political and economic dispensation to reduce the time and money it would take for the council to acquire the plots, and such powers are retrograde at a time when the government is attempting to build its image as a responsible, accountable, transparent and democratic institution with respect for private property. The adverse effects of the 1975 Land Act alluded to above should act as a warning to such a use of draconian powers. Recommendation (ii) complements (i). While it might be argued that market supply will not make land available to the poor, a public supply has also failed to meet the land needs of the poor: if the poor could easily get access to publicly supplied land, then squatting would not be necessary.

Research in Nkana-Kitwe shows that the supply of conventional housing has not responded to the needs of the poorest. The disposal of institutional housing seems to be supported by the market assumption that promoting the formal private housing market will result in more houses, greater competition resulting in greater levels of efficiency and lower costs to the public so that low wages cease to be a hindrance. Although the current formal housing situation in Zambia is one of deficit (need exceeds supply), economic conditions and housing development infrastructure are poor and cannot support a buoyant formal land and housing market. Wages are low, interest rates prohibitive, and mortgage lending limited. The sale of institutional housing gives sitting tenants a reasonable chance to participate in the formal housing market. The majority of households outside what was institutional housing are in site-and-service schemes, upgraded informal settlements, and squatter settlements. A land and housing market proposal should be designed for the bottom end, outside what is commonly accepted as the conventional housing market, where people may not meet standard requirements for mortgage finance. Unless targeted at those who have not benefited from the sale of institutional housing, the proposal is unlikely to improve access to land and housing for the poor.

Recommendation (iii) - offering government guaranteed mortgages at low interest rates: The effectiveness of this proposal is limited by the number of financial houses willing to grant credit at sub-economic rates of interest. Existing private credit houses are unlikely to accept such a proposal unless the difference is made up through a government subsidy.

While mortgage subsidies can keep the cost of land and housing low, they also may have the effect of countering the recommendation on promoting the market supply of land. Low-interest mortgages must take account of the possible effects of the subsidy on the economy: for example, taxation to fund subsidies might generate a deadweight loss, in that the tax collected fails to cover the loss incurred by allocating land and housing at a subsidized cost. Another effect is the possibility of encouraging rural-urban drift, and there is the ever present and difficult problem of identifying subsidy cheats. Unless there is reasonable economic growth, a subsidy imposes a financial burden on individuals and companies struggling on limited incomes and profits. The present Zambian economy has a shrinking industrial base and a poorly paid and diminishing workforce. For subsidies to be viable, taxes should target that sector of the economy for which a tax acts as an incentive for growth: for example, a well designed rates structure can encourage the productive use of land. With economic growth, increased wages and/or the creation of more job opportunities should improve the level of affordability and thus limit the need and level of subsidies. When urban growth has slowed down due to a number of factors, a recommendation which might once again increase urban growth should only be taken after considered analysis of the economic and social implications for urban management.

Given prevailing economic conditions, especially for the majority of the households who have not benefited from institutional house sales, the only way the proposal on a market supply of land and housing can be justified is by offering some form of subsidy. The Draft Housing Policy (GRZ, 1995b) estimates that less than ten per cent of the Zambian population can afford the cost of an urban dwelling designed to minimum standards. A land and housing market has to meet the needs of a majority of people who cannot afford even the cost of minimum urban shelter. Means-tested low-interest mortgages should be guaranteed by the government for the benefit of the poor.

With the sale of rental houses, the public role in housing is diminishing. Because a completely private land and housing market is unlikely to respond to the housing requirements of the poor, housing should move from the public domain to an alternative semi-private land and housing market, until conditions are ready for a complete private provision. The economic reality is that an immediate and complete state withdrawal in the supply of land and housing is not possible. A level of government presence, if only to mediate between conflicting economic and social goals, is necessary. Applied properly, low-interest mortgages or subsidies can be an effective tool for regulating land and housing values in the market.

Recommendation (iv) - decentralizing the land allocation process: Decentralizing land allocation and management can increase the level of access to land by bringing decision-making and registration to the people at city level. It conforms with government's policy of decentralization and meets local authorities' desire for autonomy, which is meaningless if local authorities still rely on central government funding. The withdrawal of the headlease system (Chapter 4) conflicts with a policy of decentralization and undermines local authority independence. A decision to devolve land management complements two earlier recommendations. Demand and supply are likely to respond quickly to changes in value in a decentralized system. The element of monopoly in market information which would characterize a centralized system is reduced. Market responses and the competition engendered when information is diffused can only add to market efficiency. We see therefore that decentralization of the land allocation process is complementary to a market supply of land.

So far, all the recommendations made are directed at improving the market supply of land and housing - a form of access largely dominated by affluent households. The greatest need, however, is among the majority poor who, unless subsidized, cannot afford the market value of land or housing. In the absence of any subsidy, how can these gain access to land and housing? This leads to the final two proposals.

Recommendation (v) - re-investment in public housing: The effect of the sale of local authority housing, while benefiting the sitting tenants, has been inequitable to others outside local authority housing, redistributing income to a section of the community. A suggestion is to re-invest the proceeds from such sales in building more public housing to continue such income redistribution. If local authorities still fail to re-invest in public housing, then individual initiatives to provide own housing should be supported.

Recommendation (vi) - facilitating individual housing initiatives: Proposals on land and housing subsidies and re-investment in public housing rely on the availability of public funds, without which these proposals cannot be implemented. The emergence and development of the informal housing sector however, demonstrates the ability of individual households to provide their own shelter with no help from the government.

　　While urban growth was limited, migration controlled and the economy relatively well off, colonial local authorities could afford to invest in social housing which facilitated the reproduction of labour to support capitalist industries. With rapid urban growth after independence and a worsening economy, public investment in social housing has ceased, and city growth has taken place in the informal housing sector, outside the control and without the help of public institutions. Squatter upgrading initiatives show the planners (public institutions) coming from behind with their tools (social and physical infrastructure) while the city strides on ahead. Given this reality and instead of frustrating these individual investment initiatives, local authorities should work with the informal housing sector, persuading, negotiating, advising and enabling a rational city development in which individuals invest their time and money, and public authorities offer tenure security, technical advice and infrastructure services.

Conclusion

A number of social and economic aspects of housing in Nkana-Kitwe have been explored. The mining company and local authority have been the two main providers of housing. Improvements in African housing can be seen in the shift from temporary and bachelor housing units, with small habitable rooms and limited sanitary facilities, to permanent married units. This reflects economic and moral arguments related to labour stabilization, coupled with nationalist and labour agitation for improved social facilities.

　　An analysis of the financial structure of local authority housing reveals that a segregated accounting system (the Durban System) provided for the self-financing of low-cost housing. The level of revenue collected from low-cost areas was the same as that from high-cost areas, but because of the high density at which low-cost areas had been developed, there was a greater level of cost-sharing, albeit for lower standards of services. Analysing the financial structure, the now characteristic deficit of the housing revenue account appeared first in 1964, when surpluses from liquor sales meant to finance general social welfare in low-cost areas were diverted to meet the housing deficit. Rentals failed to match general price increases: between 1958-65 and 1967-75 rentals are reported to have remained the same. Local authority attempts at housing have not met the demand, hence the move after independence to site-and-service schemes.

Although intended to address the housing needs of the poor, cost constraints have restricted the supply of and accessibility to site-and-service schemes by the poor. Where the poor managed to gain access, the sites were often re-sold or remained undeveloped. The verdict on formal housing processes, whether the provision of complete rental units or serviced sites, is one of limited success, and as a result households have opted for self-determination in informal settlements.

From almost nil in 1964, as a proportion of the entire Nkana-Kitwe population, the number of people residing in informal settlements rose to a high of 30 per cent in 1969. This has since dropped due to a number of factors such as re-settlement in site-and-service areas and other demographic factors as affected by economic and social conditions. This group constituted about 26 per cent of the population in Nkana-Kitwe in 1998. In terms of the comparative supply of dwelling units between the informal and local authority low-cost housing, the proportion has grown from almost 0:100 at independence to 100:89 in 1998.

An attempt has also been made to quantify the need for housing in Nkana-Kitwe. A comparison between the estimated current population in terms of households and the number of existing formal dwelling units gives a shortfall of 11,600 units, defining the 1998 housing need in Nkana-Kitwe.

A number of practical suggestions have aimed at addressing Nkana-Kitwe's housing problems, particularly to infill existing high-cost settlements with additional low- or medium-cost houses. Other proposals are for a controlled market supply of land and housing, support for individual housing initiatives, and the re-investment of proceeds from the sale of local authority housing into public housing.

This chapter illustrates the presence in the city of forces through whom urban management decisions on housing find expression, or whose influence bears on such decisions. These forces, if balanced, could promote a creative tension, such as that which led to the improvement of African housing in Nkana-Kitwe. While the form of housing reflects the technical nature of urban management, the housing process, dominated by tensions between mining capitalism, colonial government, labour and nationalist agitation confirms the theoretical premise about urban management being both a technical and political process on resource allocation between competing, and sometimes conflicting, interests. In Chapter 6 and with regard to plot premiums, the government has been portrayed as an interested party willing to benefit freely from the premium fund. In relation to African mine worker housing, the government played an advocatory role, pushing a reluctant mining company to improve African housing.

This chapter also provided evidence of how urban management decisions have affected distributive outcomes, largely excluding the poor and supporting the argument for strong moral principles to guide decision-making. For example, while the local authority in Kitwe retained its rental housing and only allocated this to those in formal employment or with a regular income, the poor and unemployed were excluded. When subsidies were extended to tenants of local authority housing, the benefit again went to those who were either employed or had a regular income (criteria used for allocating a local authority house) - again the poor and unemployed were excluded. The sale at reduced prices of local authority houses to sitting tenants offered a third bite at the cherry for tenants housed and subsidized by public funds, for being able to afford regular rentals. The decision to sell local authority housing is therefore seen as a redistribution of income to people already favoured to be local authority tenants in the first place. If the money from such sales was re-invested to build more local authority housing, then a greater income redistribution is possible as more and more households are availed access to public housing.

This chapter is the last of four which have attempted to support, with empirical evidence from Nkana-Kitwe, theoretical statements made in developing the central argument, and the resulting proposal for a moral and ethical framework of principles to support urban management decisions related to land, housing and public services.

The following concluding chapter brings together these findings, develops a general framework of principles and attempts a contribution towards an explanatory theory of urban management in Nkana-Kitwe and Zambia.

Chapter 9

Conclusions

Introduction

In reviewing the development of theory on urban management, a central argument is for a moral framework of principles guiding urban management decisions, which is restated in the opening section of this chapter. Based on empirical observations made on a number of identified urban management issues in Nkana-Kitwe, the weight of evidence supporting theoretical postulates made in developing the central argument is presented in the second section. The third section posits a general moral framework of urban management to guide policy on issues such as land, housing and public service provision. The fourth section focuses on the development of the structural conflict model of urban management in Zambia. The closing section identifies issues needing further research.

Restatement of the Central Argument and Methodology

Urban management is a recent approach to urban problems in the developing world. Notwithstanding its recent prominence, urban management borrows from traditional disciplines and has not yet developed its own theory.

Urban management has been shown to be both a technical and political process through which decisions about resource allocation are made. Because of its potential to affect distributive outcomes, a moral and ethical framework is proposed to guide decisions in resource allocation as the central argument of this book. Because of the problem human self-interest creates for moral and equitable decision-making, the structural conflict model proposes the maintenance, in balance, of existing tensions in urban areas so as to facilitate the adoption of mutually beneficial decisions.

An eclectic approach encompassing interpretivism and structuralism, both drawing on empirical evidence, has been used in the methodology design of this research, using observable evidence to support the central argument and to illustrate the positive outcomes of the proposed structural conflict model of urban management.

The following section examines the weight of evidence in support of the central argument, and the need for a moral and ethical framework to guide urban management policy decisions.

Research Findings

The focus of this research has been to explore the distributive functions of urban management between competing and sometimes conflicting interests. It has examined how power relationships in the development of Nkana-Kitwe have influenced its built form and urban administration. Having identified the main interest groups in the historical development of Nkana-Kitwe, it has explored a range of urban management issues relating to the distribution and taxation of land and property; provision and financing of services, cost recovery; and finally the form of and agencies in housing, and the financial structure of local authority housing. What follows is a re-statement of the findings and how they relate to the introductory argument.

In tracing the origins of Nkana-Kitwe in Chapter 5, two dominant forces have been identified: economic (shared between the mining company, small private capital and the colonial government), and political (dominated by the colonial government). An examination of the city's built form reveals that both the mining company and the Territorial government established segregated townships, revealing the value judgements about the kind of society each wanted to create.

In analysing the continuity of interest groups and spatial form, I have shown that political and local authority executive (technical) powers have had a dominant role in shaping Kitwe's development after independence, replacing the small private capital as represented by the mostly settler community of traders who had contributed significantly to the establishment and management of Kitwe. Despite attempts at integration, the mining company still exerts a strong influence on Kitwe, and exists as an administrative authority in Nkana. Since independence, the ability to influence Kitwe's development is no longer the preserve of formal power structures, but extends to informal power bases such as squatter settlements.

On the continuity of physical form, evidence shows that the linear structure and segregated residential structure of the city core has been maintained, although since independence segregation is not on racial basis but social. In addition, the siting of some of Kitwe's site-and-service schemes after independence is largely dictated by the siting of the intended beneficiaries, which means informal settlements. While topography and segregation led to fragmented development even in the city core, the lack of morphological unity, especially in Nkana-Kitwe's peripheral development, seems to have continued after independence. In summary, the focus on power play in Nkana-Kitwe's development has established the existence of, and shifts in, a multiplicity of power bases which influenced the development of the city's form and administrative structures.

Having established the existence of interest groups in Nkana-Kitwe, the discussion examined for specific issues the relationships between these groups focusing on the way in which urban management has allocated burdens and benefits. Chapter 6 examined the supply of land and property in general, rate taxes and the provision of public services funded from rates. In examining the mode of supply through which institutions and individuals gain access to land and property, it has been established that, although the state has had a monopoly in the ownership of land, and notwithstanding government controls on the land and property market instituted in 1975, the market (in leaseholds) has had a significant share in the allocation process. However, in comparing the population levels of formal and informal housing areas in Chapter 8, it has been shown that both the formal market and the government failed to supply land and housing to about 26 per cent of the population in Nkana-Kitwe in 1998.

Evidence from the spatial and temporal analysis of land and property values revealed that the presence of political (planning control) and economic forces (mining company) have influenced what approximates to the classical value gradient theory.

An analysis of the financial arrangements for the general rating system, and the resulting relationships between institutions and/or individuals in Kitwe, reveals a segregated 'benefactor pays' structure, through which Africans and Europeans paid for their own services. In accord with the country's socialist policy, the segregated financial structure was replaced with the 'one town' concept in 1966 with the objective of creating an integrated society, a decision with implications on the distributive outcomes in urban management.

Evident once again in this chapter is the existence of self-interested groups, shown by the acrimony in the relationships between the local authority, central government and the mining company. Government's self-interest has been shown by its preparedness to benefit from a fund to which it consistently refused to contribute. In relation to the premium fund, this evidence dispels the notion of 'public interest' and reveals that, instead of mediating between conflicting interests, the government is itself an interested party, hence the need for a balance of forces with the potential to create just relationships in the distributive outcomes of urban management. While the government as an institution behaved in a selfish manner in the matter of premium funds, I have been able to show the advocatory role played by individuals in government on behalf of the local authority in Kitwe.

Chapter 7 examined the issue of public services with a view to establish the institutional and physical structure of water supply and to explore the question of equity. It has been shown that the mining company and local authority co-operate to supply water to an estimated 91 per cent of the city's population, with higher service options concentrated in the historically European or high-cost residential areas. Detailed estimates of consumption levels after 1974 have led to the conclusion that, while consumption rose rapidly between 1955-74, after 1974, and despite the growth in the city, constraints in production and distribution have limited the supply of water. It has been estimated that 1990 consumption levels fell below their 1974 levels. Estimated domestic consumption levels in Nkana-Kitwe are generally high and, if moderated, the existing capacity could supply an additional one and a half times the number of existing domestic consumers. However, a comparison between income and water charges suggests that, even if it was possible to moderate consumption and thus redistribute water to a wider population, affordability would still be a problem. Analysing the financing of water supply in relation to equity and economic efficiency in cost recovery, Kitwe's local authority objective in fixing water charges is to match revenue with expenditure while cross-subsidizing consumption in low-cost residential areas in which fixed charges are levied irrespective of the level of consumption. Regarding the tariff structure, two observations made are that it rewards high domestic consumption, and that the grouping of commercial and industrial use under one consumer group is uneconomical (in that it allocates an unusually high initial volume to commercial users).

A number of factors have been advanced as contributing to the problem of cost recovery today, including: a mismatch between the high service options designed for a high-income European population and the historically low salary structures for Africans who have inherited the city; the colonial non-profit objective for services to African areas reflected in low tariff structures which have been difficult to adjust upwards; political interference hindering the application of sanctions to defaulting consumers; and the local authority's own system inefficiency.

Analysis of the different pricing policies established that because of the existing level of subsidy in which a consumer on a meter is expected to subsidize another 1.6 fixed rate consumers, the overall identity of the consumer population who stand to gain or lose by a change in pricing policy would remain the same. On balance, however, the choice fell on the total average cost, because of its potential to improve cost recovery and offer a cross-subsidy to poor households on the periphery. In arriving at the total average cost, questions of equity, morality, affordability and willingness to pay had to be balanced against other economic objectives necessary to ensure a sustainable service.

The current tariff structure, although offering a substantial subsidy to fixed rate consumers in low-cost houses, is considered to constitute a heavy charge on those metered households expected to cross-subsidize households on fixed rate consumption, and also to fail to limit subsidized consumption. A further proposal is to impose a limit on subsidized consumption to the basic need, in an attempt to arrive at a balance between equity and economic efficiency on one hand and, on the other, the social objective of subsidizing only that level of consumption necessary to meet the basic needs for a flourishing community in which part, if not the full cost, is paid by the respective consumer population. This chapter highlighted both the technical and political process of urban management, and showed how decisions on water supply have affected not only the physical distribution of service levels, but also the distribution of burdens and benefits in the tariff.

Housing, discussed in Chapter 8, is another central theme in the discourse on urbanization in Zambia. The institutional structure in the provision of housing shows yet again the predominance of the mining company and the local authority. After observing general improvements in the form of housing, economic and moral arguments for the stabilization of labour alongside nationalist and labour demands for improved social facilities have been identified as catalysts to this progress.

Tensions between mining capitalism, colonial government and the agitation from the general populace organized as labour or political movements provided yet more evidence of the multiplicity of forces and the creative nature of these tensions in housing development in Nkana-Kitwe.

In Chapter 6, the government is presented as self-interested. But government's ability to challenge the mining company in the matter of African housing and to allow itself to address, through the local authority, the issue of African housing, is evidence either of an awareness by the government and the mining company to pacify nationalist concerns so as to head off political activity and ensure social stability, necessary for the continued exploitation of the mineral resources or a genuine concern for the advancement of the general African population.

With respect to the segregated accounting system first referred to in Chapter 6, a similar principle in housing provided for the self-financing of low- and high-cost housing, with a greater level of cost-sharing in low-cost high-density areas.

The now characteristic deficit in the local authority's housing account, first shown in 1964, reveals a problem of cost recovery, again the result of low rental structures which have not been adjusted to correspond with general price increases partly because of poor salary structures. For example, when the deficit first came up in 1964, instead of raising rentals to economic levels, and despite government urging to do so, the local authority opted to transfer surpluses from liquor sales to the housing account and to ask for a government subsidy in order to cushion the adverse impact economic rentals would have on poor households. The general observation on the formal housing process, either in the form of complete rental units or serviced sites, is that it failed to meet the demand and needs of the poor, leading to the emergence and growth of informal housing units from almost zero at independence, to about 35 per cent of all formal housing units in Nkana-Kitwe in 1998, or contributing a proportion of 100 units for every 89 low-cost units provided by the local authority in 1998.

Using a number of detailed direct and indirect measures to determine relationships between the cost of land, housing and finance, the crucial aspect hindering the supply of and access to formal land and housing is shown to be financial. It is proposed at the end of Chapter 8 that a process of densification might provide additional serviced sites in existing low-density areas. Other proposals are for a controlled market supply of land and housing, support for individual housing initiatives and for re-investment in public housing.

By looking at the distributive outcomes of decisions made in public housing (allocation policy, subsidies and the sale of public housing), this study illustrates the resulting inequity at each stage, and provides further arguments for a moral framework to guide policy in Zambia's urban management.

General Policy Recommendations

This research has established the existence of poles of interest in the city and demonstrated the creative potential engendered by their presence. With respect to land, housing and public services, allocative outcomes have also been shown to reflect urban management decisions modified by the influence of existing forces. In order to ensure that urban management policy provides for an equitable distribution of services and benefits and for a redistribution to poorer groups, a set of moral and ethical principles to inform choices in urban management in Zambia are proposed.

Urban policy should promote economic growth: A recurring theme associated with urban problems in Nkana-Kitwe and Zambia is finance. Financial constraints have hindered the supply of and access to land and housing, and also the sustainable provision of public services. Urban policies, therefore, should aim to stimulate economic growth. In a labour surplus country such as Zambia, economic growth and more job opportunities will not only endear the government to the people, but also widen the tax base to fund public services, and even improve cost recovery for urban services. Only as the economy grows will the country be able to tackle urban social issues. Economic growth is thus an essential corollary to redistribution.

Urban policy should foster just social relations: Because urban management decisions affect resource distribution, equitable and moral principles should guide such decisions if the distribution is to be fair and just. For a private good such as water, the principle of 'benefactor pays' is equitable, inculcates a sense of responsibility, and challenges free riding. It also empowers and promotes the dignity of the community concerned by freeing the people from vulnerability to political manipulation and the spectre of poverty which has often been associated with the urban poor. To deny basic services to those unable to pay for them is immoral, so there should be a balance between rights and responsibilities. Those with rights to urban services because they have paid for them might have to forego the exclusive right of enjoyment in order to allow for a redistribution to poorer groups, who should also pay a little towards the cost. Responsibility might initially need to be enforced, and eventually learned and acquired by successive generations. The initial imbalance in the distribution of resources and the threat to social stability is one of the motivating factors to redistribution. The principle of 'benefactor pays' reinforces responsibility for those benefiting from a redistribution of benefits.

Urban policy should be community centred but...: The inherited segregated structure of Zambia's urban areas, in Nkana-Kitwe exacerbated by the twin town structure, militates against the physical unity of urban communities, so efforts should be made to promote a corporate identity based on other shared values and interests. By community, it is not necessarily intended to mean the integration of housing across social strata, although this might be a possibility. It is equally possible to have a community spanning across territory, a wider identity and unity enriched by the social diversity across communities but bound by some common bond. Unless residents feel part of a community and participate in its management, it will be difficult to encourage that sense of responsibility and accountability related to civic duties and social relationships. For example, the redistribution of benefits through cross-subsidies from one community to another is easier when the subsidizing community feels a sense of relationship with the recipient community.

Urban policy should respect the individuality of every household: To the extent that the concept of community can be encouraged, urban policy should aim to promote that level of integration which can build strong and coherent communities. Every household should be enabled to own a home within which secure and exclusive use rights can be enjoyed. The ownership of such property has the potential to promote family stability necessary for a strong community, go towards instilling a sense of pride and dignity necessary for self-actualization, and raise the degree of morale and co-operation in civic duties arising from the stake in the community which such ownership confers. The enjoyment of individual rights is likely to be limited by the threat of social instability from those that do not have this stake, so urban policy should ensure, as far as practicable, the inclusive allocation of resources.

The above general recommendations explicitly incorporate an element of moral values to guide urban management at policy level, and depend upon state support. Yet why should the state apparatus of politicians and civil servants, with their own interests to serve, implement policy which could redistribute resources at their expense? This question leads to the proposed structural conflict model of urban management.

A Structural Conflict Model of Urban Management

Urban management is both a technical and political process through which resources are distributed to competing groups, whose composition varies. Primarily political and economic interests dominate but, as has been shown in this research (Chapters 4 and 5), power is not the preserve of politicians and capital - sections of society can rally around a common interest and influence decisions. Given that self-interest (whether the self is a corporate body or an individual) generally determines how these interest groups behave, can they be persuaded to act in ways benefiting not only themselves but others?

The experience of Kitwe is informative at this point. Kitwe public township was developed from private finances; the mining company gave up land it had leased from the Territorial government, and offered its municipal services for the establishment of a public township. Individual traders offered capital to the government for the development of Kitwe. We therefore see Kitwe develop after a time of considerable tension between the mining company's economic interests and the Territorial government's political and economic interests. The tension in this case was constructive not destructive. Capital is willing to invest in the community if it thinks that its operations are dependent on that community.

Further conflict between the Territorial government, mining company, organized labour and nationalist sentiments led to the improvement of housing conditions for the Africans. After nationalization there was no real threat to mining interests, since these were vested in the central government, which means that the local authority cannot use political clout to get the mines to contribute to the development of the town. Falling more under central government (at least until the ongoing re-privatization process is over), the mining company does not need to be good corporate citizens. Although there is an understanding that the mine sells water to the local authority and this is offset by the adjustment of rates on mine property, there have been constant wrangles in which the council always came off the worse. Notice therefore the negative effects of nationalization which removed the strong group interests between the government and the mines, a tension which had been harnessed for the good of the town.

The reference so far has been to two power bases, economic and political, but the city is a complex entity containing other interest groups. Perhaps the next group we will consider is the 'people', a conglomerate of individuals, communities and organized groups. Important organized groups in the 'people' are the Church, organized labour and civic groups. The main power bases are reduced to three, thus: the state, capital and the people.

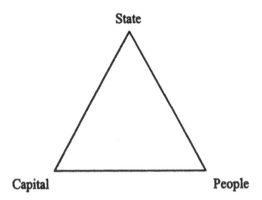

Source: Author.

Figure 9.1 Power bases in urban management

The success of this approach depends upon the relative balance of power between interest groups: no one group should become too strong so as to upset the balance. For a country in transition such as Zambia, it might mean an incremental change from a centrally-planned economy to a market economy to arrive at a balance of power between private capital and the state. A complete shift to a market economy, given current low levels of political articulation and the country's labour surplus, might make private capital too strong, and government incapable of being the counter-balancing force to its greed and excesses. Although it has its own interests to serve, the state is a major player in the structural conflict model. For example, I have been able to show that in the case of African mine-worker housing, the government was able to play an advocatory role. While it might need to be more accountable, attempts at a more progressive urban policy before and after independence show the potential for a responsive state: the improvement in housing, universal franchise in local authority, squatter upgrading policies, attempts at societal integration through physical planning and fiscal policy, and the sale of council housing to empower people.

The effect of an imbalance in power relations is shown by the experience of Nkana-Kitwe and Zambia. Unfulfilled expectations resulted in nationalist uprising and eventually independence. The tension in this case is both destructive to the old colonial order and constructive of a new post-colonial order with its new tensions.

Given existing imbalances in the distribution of power, the ongoing privatization of business interests and institutional housing offers an opportunity to ensure that power is diffused more widely in the country. As for the disadvantage suffered by the majority of the population - a history of marginality and inequality leading to a low level of political articulation - it is important that these people are helped to articulate their needs. While the structural conflict model rests on the presence of self-serving interests, I also concede that government and other organizational and individual efforts can further the cause of others, either because of the common gains from such an advocacy or a general concern for social justice. The reconciliation of competing interests requires that power is widely dispersed.

This concept is based upon the understanding of the true and selfish nature of humanity and the city as a domain of conflict; it makes no pretensions about the true nature of humanity, level of communality or integration, but recognizes that society is a shared endeavour. Thus the concept has a certain moral coherence – it does not assume progression to complete harmony and equality, but allows resolution of certain conflict points to create conflict in others situated at another level. In Nkana-Kitwe, power bases and tensions shift before and after independence.

The basis of the structural conflict model of urban management is the acceptance of the self-serving nature of humanity and the dialectical character of society as a unity of competing but dependent interests all grounded in the empirical observation of the progress made in urban policy as a result of tensions between group interests.

How does a structural conflict model relate to a moral and ethical framework for urban policy? The model does not introduce conflict, but recognizes the existence of competing and potentially destructive forces. With proper management, these forces can provide conditions which foster the implementation of urban policy and take account of the moral and ethical concerns in the distributive outcomes. The model redirects energies with a destructive potential, and helps society deal with conflicts in resource distribution through those tensions which could escalate conflict and violence.

Further Research Directions

While it has been possible to establish the constructive effects of a multiplicity of forces in Nkana-Kitwe, and the distributive influence of urban management in resource allocation, some areas justify further examination and theorizing:

(i) although the structural conflict model depends on a relative balance of power, no investigation or theoretical suggestions on the methodology and techniques involved in arriving and maintaining such a balance from the present position of imbalance have been offered;

(ii) to elaborate on the changing dynamics of Zambia's urban areas, and how these affect the conflicts and the composition of interest groups; and

(iii) to what extent distributive outcomes are influenced by, or independent of, group interests, and how power is distributed and used to influence urban policy.

Bibliography

Akeroyd, A.V. and Hill, C.R. (eds.) (1978), *Southern Africa Research in Progress*, Collected Papers 4, Centre for Southern Africa Studies, University of York.

Allen, J.B. (1966), *The Company Town in the American West*, University of Oklahoma Press, Oklahoma.

Almond, G. (1993), 'Foreword: The Return to Political Culture', in L. Diamond (ed.), *Political Culture and Democracy in Developing Countries*, Lynne Rienner, London, p. xi.

Altaf, M.A. and Hughes, J.A. (1994), 'Measuring the Demand for Improved Urban Sanitation Services: Results of a Contingent Valuation Study in Ouagadougou, Burkina Faso', *Urban Studies*, Vol. 31(10), pp. 1763-76.

Alterman, R. (ed.), (1988), *Private Supply of Public Services*, New York University Press, New York.

Amin, S. (1969), 'Le monde des affaires Senegalais', in W.J. Hanna and J.L. Hanna (1981), *Urban Dynamics in Black Africa: An Interdisciplinary Approach*, 2nd edn., Aldine Publishing Company, New York.

Amos, J. (1993), 'Planning and Managing Urban Services', in N. Devas and C. Rakodi (eds.), *Managing Fast Growing Cities*, Longman, Harlow, pp. 132-52.

Angel, S. *et al.* (eds.) (1983), *Land for Housing the Poor*, Select Books, Singapore.

Angel, S. *et al.* (1987), *The Land and Housing Markets of Bangkok: Strategies for Public Sector Participation*, PADCO, Washington, DC.

Angelo, S. (1983), 'Upgrading Slum Infrastructure: Divergent Objectives in Search of a Consensus', *Third World Planning Review*, Vol. 5(1), pp. 5-22.

Anon (no date), *General Information on the Town of Kitwe*, filed at the Zambia High Commission Library, London.

Anon (1953), 'Civic Association may Request Full Investigation of Plot Premiums', *Staff Reporter*, 28 September, filed at ZNA: SEC 1/1528.

Anon (c.1959), *Africans of Nkana*, Rhokana Corporation Ltd.

Anon (1971), 'Squatter Problem', paper presented to the Town Clerks' Society of Zambia by Kitwe City Council's Town Clerk.

Anon (1993), 'Cornerstone of UNIP's Land Policy', paper presented to the National Conference on Land Policy and Legal Reform, July 1993, Freedom House, Lusaka.

Ashcroft, J. and Schluter, M. (1989), *Christian Principles for Urban Policy*, The Jubilee Centre, The Newick Park Initiative, Cambridge.

Balchin, P.N., Bull, G.H. and Kieve, J.L. (1988), *Urban Land Economics and Public Policy*, 4th edn., Macmillan Education Ltd., Basingstoke.

Balchin, P.N. and Kieve, J.L. (1985), *Urban Land Economics and Public Policy*, 3rd edn., Macmillan Education Ltd., Basingstoke.

Baloyi, D.R. (1983), *The Role of Medium-Sized and Small Towns in Zambia: 1969-1980*, University of Aston, Birmingham.

Bamberger, M., Sanyal, B. and Valverde, N. (1982), *Evaluation of Sites and Services Projects. The Experience From Lusaka. Zambia*, The World Bank Staff Working Paper, No. 548, Washington.

Bancroft, J.A. (1961), *Mining in Northern Rhodesia: A Chronicle of Mineral Exploration and Mining Development*, The British South Africa company.

Banfield, E.C. (1964) 'Politics', in J. Gould and W.L. Kolb (eds.), *A Dictionary of the Social Sciences*, Tavistock Publications, London, pp. 515-17.

Barlowe, R. (1986), *Land Resource Economics: The Economics of Real Estate*, 4th edn., Prentice-Hall, London.

Baross, P. and van der Linden, J. (eds.) (1990), *The Transformation of the Land Supply Systems in Third World Cities*, Avebury, Aldershot.

Bate, T. (1994), 'An Introduction to Rates', paper presented to the Council Rating Workshop, 10-11 December 1994, Andrews Motel, Lusaka, Zambia.

Bateman, A.M. (1931), *Mining and Metallurgy*, July, New York.

Batley, R. (1993), 'Political Control of Urban Planning and Management', in N. Devas and C. Rakodi (eds.), *Managing Fast Growing Cities*, Longman, Harlow, pp. 176-206.

Batley, R. (1996), 'Public-Private Relationships and Performance in Service Provision', *Urban Studies*, Vol. 33(4-5), pp. 723-51.

Beatley, T. (1984), 'Applying Moral Principles to Growth Management', *Journal of the American Planning Association*, Vol. 50, pp. 456-69.

Beatley, T. (1991), 'A set of Ethical Principles to Guide Land Use Policy', *Land Use Policy*, January, pp. 3-8.

Beatty, C. (1931), *Mining Metallurgy*, December, New York.

Behrens, R. and Watson, V. (1996), *Making Urban Places - Principles and Guidelines for Layout Planning*, Urban Problems Research Unit, University of Cape Town.

Bostock, M. and Harvey, C. (eds.) (1972), *Economic Independence and Zambian Copper*, Praeger, New York.

Bower, P., Christiansen, S. and Howard, J. (1997), *Informal Settlements and Security of Tenure: Some Lessons From Zambia and Application to Namibia*, iKUSASA, Durban, South Africa, 24-28 August.

Bruce, J., Kachamba, F. and Hansungule, M. (1995), 'Land Administration: Processes and Constraints', in M. Roth (ed.), *Land Tenure, Land Markets and Institutional Transformation in Zambia*, research paper, Land Tenure Centre, University of Wisconsin-Madison, pp. 47-77.

Bubba, N. and Lamba, D. (1991), 'Local Government in Kenya', *Environment and Urbanization*, Vol. 3(1), April, pp. 37-59.

Bulawayo Chronicle (1947), 'Kitwe has Developed on Novel Lines', *Bulawayo Chronicle*, 26 September, filed at ZNA: SEC 1/1528.

Bull, T. (1992), 'Multiplication by Division', *Profit*, October, pp. 11-12.

Bull, T. and Simpson, D. (1993), 'ZCCM - Privatization's Golden Opportunity', *Profit*, March, pp. 8-16.

Bullard, R. (1993), "Tenere - to Hold", paper presented at the Conference on Land Tenure, December, RICS and University of East London, Duncan House, UK.

Caincross, S. (1990), 'Water Supply and the Urban Poor', in S. Caincross *et al.* (eds.), *The Poor Die Young: Housing and Health in Third World Cities*, Earthscan Publications, London, pp. 109-26.

Caincross, S., Hardoy, J.E. and Satterthwaite, D. (eds.) (1990), *The Poor Die Young: Housing and Health in Third World Cities*, Earthscan Publications, London.

Carney, M. (1992), 'The Widening Role of the Water Companies. Proceedings of the Institution of Civil Engineers', *Municipal Engineer*, Vol. 93(4), pp. 211-14.

CASLE (1981), *The Role of the Surveyor in Developing Countries with Particular Reference to Central and Southern Africa*, report of proceedings, Malawi, April.

CASLE (1983), *The Development of Africa Land Resources with Particular Reference to Central and Southern Africa*, report of proceedings, Zimbabwe, September.

Castells, M. (1977), *The Urban Question: A Marxist Approach*, Edward Arnold, London.

Chanda, A.W. (1994), *Report of the Proceedings of the Land Bill Seminar Held at Pamodzi Hotel*, 23 September, University of Zambia, Lusaka, Zambia.

Cheema, G.S. (ed.) (1984), *Managing Urban Development: Services For the Poor*, United Nations Centre for Regional Development, Nagoya.

Cheema, G.S. and Rondinelli, D.A. (eds.) (1983), *Decentralisation and Development*, Sage Publication, London.

Cherry, G.E. (ed.) (1980), *Shaping an Urban World*, Mansell, London.

Chileshe, J.H. (1987), *Third World Countries and Development Options*, Vikas Publishing House Pvt. Ltd., Zambia.

Chileshe, R.A. (1991), 'Land Development Policies - The Need For Change', report on the seminar held by the SIZ at Pamodzi Hotel, Lusaka, November, School of Environmental Studies, Copperbelt University, Kitwe, Zambia.

Chitoshi, C.M. (1984), 'Suggestions for Improving the Financing of Local Administration in Zambia', *Planning and Administration*, Vol. 11(2), pp. 16-22.

Chiwele, D.K. (1992), 'How Zambian Households Survive the Economic Crisis', *The Weekly Post*, 3-9 July.

Christopher, A.J. (1988), *The British Empire at its Zenith*, Croom Helm, USA.

City of Kitwe (c.1967), *Quinquennial Review*, filed at KCC, DS/l03/57/2 as Annexure A to Report No. 112.

City of Kitwe (1970), *Four Years of Progress*, Kitwe City Council, Kitwe, Zambia.

City of Kitwe, *City of Kitwe Street Plan*, published by Survey Department, Lusaka.

City of Kitwe, *City of Kitwe Development Plan, 1975-2000, Survey of Existing Conditions*, NHA, Lusaka.

City of Kitwe, *City of Kitwe Development Plan, 1975-2000, Final Report*, NHA, Lusaka.

Clarke, G. (1991), 'Urban Management in Developing Countries - a Critical Role', *Cities*, May, pp. 93 -107.

Clay, G.C.R. (1949), 'African Urban Advisory Councils on the Copperbelt of Northern Rhodesia', *Journal of African Administration*, Vol. 1, pp. 33-8.

Collins, J. (1969), 'Lusaka: The Myth of the Garden City', in D.H. Davies (ed.), *Zambian Urban Studies, No. 2*, Institute for Economic and Social Research, University of Zambia, Lusaka, pp. 1-32.

Collins, J. (1978), 'Home Ownership Aspects of Low Cost Housing in Lusaka', in G.W. Kanyeihamba and J.P.W.B. McAuslan (eds.), *Urban Legal Problems in Eastern Africa*, Scandinavian Institute of African Studies, Uppsala, pp. 104-25.

Collins, J. (1980), 'Lusaka : Urban Planning in a British Colony. 1931-64', in G.E. Cherry (ed.), *Shaping an Urban World*, Mansell, London, pp. 227-41.

Conyngham, L.D. (1951), 'African Towns in Northern Rhodesia', *Journal of African Administration*, Vol. 3, pp. 113-17.

Cook, P. and Kirkpatrick, C. (eds.) (1988), *Privatisation in Less Developed Countries*, Harvester Wheatsheaf, Brighton.

Copperbelt Development Plan (no date).

Copperbelt University Shelter Study Team (1995), *Best Practice for Improved Environmental Conditions: Neighbourhood No. 9. Kawama, Kitwe, Zambia*, Copperbelt University, Kitwe, Zambia.

Coser, L.A. (1968), 'Conflict: Social Aspects', in D.L. Sills (ed.), *International Encyclopaedia of Social Sciences*, Vol. 3, Macmillan, New York, pp. 232-36.

C.S.O. (1971), *Statistical Year Book*, Lusaka.

Cotton, A. and Franceys, R. (1993), 'Infrastructure for the Urban Poor in Developing Countries', *Municipal Engineer*, Vol. 98(3), pp. 129-38.

Cotton, A.P. and Tayler, W.K. (1994), 'Community Management of Urban Infrastructure in Developing Countries', *Municipal Engineer*, 103(4), pp. 215-24.

Dale, P. (1995), *The Hotline Lecture 1995 – Surveying With 2020 Foresight*, Department of Photogrammetry and Surveying, University College, London.

Dare, P. and Mutale, E. (1997), 'The Readjustment and Analysis of Zambia's Isoka Primary Network', *Survey Review*, Vol. 34(264), pp. 123-31.

Davidson, F. and Nientied, P. (1991), 'Urban Management', *Cities*, Vol. 8(2), pp. 82-6.

Davies, D.H. (ed.) (1969), *Zambian Urban Studies, No. 2*, Institute for Economic and Social Research, University of Zambia, Lusaka.

Davies, D.H. (ed.) (1971), *Zambian Urban Studies No. 3*, Institute for African Studies, University of Zambia, Lusaka.

Davies, D.H. (1972), *Lusaka: From Colonial to Independent Capital. Proceedings of the Geographical Association of Rhodesia*, Geographical Association of Rhodesia, pp. 14-20.

Davis, J.M. (ed.) (1933), *Modem Industry and the African*, Macmillan, London.

Deassis, N.C. and Yikona, S.M. (eds.) (1994), *The Quest for an Enabling Environment for Development in Zambia*, Mission Press, Ndola, Zambia.

De Blij, H.J. (1968), *Mombasa. An African City*, North Western University Press, Evanston.

De Bruijne, G.A. (1985), 'The Colonial City and the Post-Colonial World', in R. Ross and G.J. Telkamp (eds.), *Colonial Cities: Essays on Urbanism in a Colonial Context*, Martin Nijhoff Publishers, Dordrecht, pp. 231-43.

Demissie, F. (1998), 'In the shadow of the Gold Mines: Migrancy and Mine Housing in South Africa', *Housing Studies*, Vol. 13(4), pp. 445-69.

Denman, D.R. (1964), *Land in the Market*, Horbat Paper No. 30, December.

Denman, D.R. and Prodano, S. (1972), *Land Use: an Introduction to Proprietary Land Use Analysis*, George Allen and Unwin Limited, London.

Denzin, N.K. and Lincoln, Y.S. (eds.) (1994), *Handbook of Qualitative Research*, Sage Publications, London.

De Solo, H. (1989), *The Other Path*, Harper and Row, New York.

De Solo, H. (1993), 'Property and Prosperity, the Future Surveyed'. Supplement in *The Economist*, 11-17 September, Vol. 32(7828), pp. 8-10.

Devas, N. and Rakodi, C. (eds.) (1993), *Managing Fast Growing Cities*, Longman, Harlow.

Dewar, D., Todes, A. and Watson, V. (1982), *Urbanisation: Responses and Policies in Four Case Studies - Kenya, Zambia, Tanzania and Zimbabwe*, Urban Problems Research Unit, University of Cape Town, South Africa.

Diamond, L. (ed.) (1993), *Political Culture and Democracy in Developing Countries*, Lynne Rienner, Boulder and London.

Doebele, W.A. (1983), 'Concepts of Urban Land Tenure', in H.B. Dunkerley (ed.), *Urban Land Policy Issues and Opportunities*, published for the World Bank by Oxford University Press, Oxford, pp. 63-107.

Dowall, D.E. (1991), *The Land Market Assessment. A New Tool for Urban Management*, joint UNDP/World Bank/UNCHS Urban Management Programme Publication, Washington, DC.

Downing, P.B. and Gustely, R.D. (1977), 'The Public Service Costs of Alternative Development Patterns: A Review of the Evidence', in P.B. Downing (ed.), *Local Service Pricing Policies and Their Effect on Urban Spatial Structure*, University of British Columbia Press, Vancouver, pp. 63-86.

Downing, P.B. (ed.) (1977), *Local Service Pricing Policies and Their Effect on Urban Spatial Structure*, University of British Columbia Press, Vancouver.

Drakakis-Smith, D.W. (1987), *The Third World City*, Routledge, London.

Dunkerley, H.B. (ed.) (1983), *Urban Land Policy Issues and Opportunities*, published for the World Bank by Oxford University Press, Oxford.

Durand-Lasserve, A. (1990), 'Articulation Between Formal and Informal Land Markets in Cities in Developing Countries: Issues and Trends', in P. Baross and J. van der Linden (eds.), *The Transformation of the Land Supply Systems in Third World Cities*, Avebury, Aldershot, pp. 37-56.

Durand-Lasserve, A. (1996), *Regularisation and Integration of Irregular Settlements: Lessons From Experience*, Urban Management Working Paper 6, UNDP/World Bank/UNCHS, Washington, DC.

Enul, J. (1967), *The Political Illusion*, Alfred A. Knopf, New York.

Engels, F. (1844), *The Conditions of the Working Class in England*, Panther Books Ltd., London.

Engels, F. (1872), *The Housing Question*, Progress, Moscow.

Environment and Urbanization (1991), 'Rethinking Local Government - Views From the Third World', *Environment and Urbanization*, Vol. 3(1), April, pp. 3-8.

Environment and Urbanization (1994), 'Basic Services: New Approaches; New Partnerships?', *Environment and Urbanization*, Vol. 6(2), pp.3-8.

Epstein, A.L. (1951), 'Urban Native Courts on the Northern Rhodesian Copperbelt', *Journal of African Administration*, Vol. 3.

Essex and Suffolk Water (1997), 'An On-going Challenge for the Leak Seekers', *Water Lines*, November, p. 2.

Fairweather, W.G. (1931a), *Land Tenure in Northern Rhodesia. Report of Proceedings on the Conference of Empire Survey Officers*, HMSO, Colonial Office, London, pp. 323-32.

Fairweather, W.G. (1931b), *Registration in Northern Rhodesia. Report of Proceedings on the Conference of Empire Survey Officers*, HMSO, Colonial Office, London, pp. 333-38.

Farvacque, C. and McAuslan, P. (1992), *Reforming Urban Land Policies and Institutions in Developing Countries*, Urban Management Programme, The World Bank, Washington, DC.

Fawcett, C.B. (1933), *A Political Geography of the British Empire*, University of London Press, London.

Financial Mail (1994), *Employers' Group Launches Drive for Workers' Homes*, supplement of the *Zambia Daily Mail*, September – October.

Fourie, c. (no date), *The Role of Local Land Administrators and Land Managers in Decentralisation, Land Delivery, Registration and Information Management in Developing Countries*, Department of Surveying and Mapping, University of Natal, Durban, South Africa.

Friedmann, J. (1992), *Empowerment – The Politics of Alternative Development*, Blackwell Publishers, Cambridge, MA.

Friedmann, J. and Wulff, R. (1975), *The Urban Transition: Comparative Studies of Newly Developing Societies*, Edward Arnold, London.

Gann, L.H. and Duignan, P. (eds.) (1970), *Colonialism in Africa 1870 – 1960*, Vol. 2, Cambridge University Press, Cambridge.

Gardiner, J.(1971), 'Some Aspects of the Establishment of Towns in Zambia During the Nineteen Twenties and Thirties', in D.H. Davies (ed.), *Zambian Urban Studies No. 3*, Institute for African Studies, University of Zambia, Lusaka, pp. 1-32.

Gibb and Partners (1971), *The Copperbelt Water Resources Survey. Final Report*, Gibb and Partners.

Gilbert, A. (1992), 'Third World Cities: Housing Infrastructure and Servicing', *Urban Studies*, Vol. 29(3/4), pp. 435-60.

Gilbert, A. and Ward, P.M. (1985), *Housing, the State and the Poor: Policy and Practice in Three American Cities*, Cambridge University Press, Cambridge.

Gluckman, M., Mitchell, J.C. and Barnes, J.A. (1949), 'The Village Headman in British Central Africa', *'Africa'. Journal of International African Institute*, Vol. XIX(2), pp. 89-106.

Gould, J. and Kolb, W.L. (eds.) (1964), *A Dictionary of the Social Sciences*, Tavistock Publications, London.

Green, K. (1993), *Land Law*, 2nd edn., Macmillan Press, London.

Greenwood, A. and Howell, J. (1980), 'Urban Local Authorities', in W. Tordoff (ed.), *Administration in Zambia*. Manchester University Press, Manchester, pp. 162-84.

Grimes, O.F. (1977), 'Urban Land and Public Policy: Social Appropriation of Betterment', in P.B. Downing (ed.), *Local Service Pricing Policies and Their Effect on Urban Spatial Structure*, University of British Columbia Press, Vancouver, pp. 360-432.

GRZ (1965), *African Housing Board Annual Report of 1965*, Government Printer, Lusaka. Shelved at ZNA : SEC 1/1568, Lusaka.

GRZ (1966a), *African Housing Board Annual Report of 1966*, Government Printer, Lusaka. Shelved at ZNA : SEC 1/1568, Lusaka.

GRZ (1966b), *Aided Self-help Housing on Site and Service Schemes*, Circular 59/66, Ministry of Local Government and Housing, Lusaka.

GRZ (1966c), *Commission of Inquiry into the Mining Industry*, Government Printer, Lusaka,

GRZ (1967), *Report of the Land Commission*, Government Printer, Lusaka.

GRZ (1968a), *Resettlement of Squatters*, Circular 29/68, Ministry of Local Government and Housing, Lusaka.

GRZ (1968b), *Aided Self-Help Housing*, Circular 30/68, Ministry of Local Government and Housing, Lusaka.

GRZ (1971), *Second National Development Plan 1972-76*, Government Printer, Lusaka.

GRZ (1987), *District Councils Revenue and Capital Estimates*, Government Printer, Lusaka.

GRZ (1988), *Ministry of Decentralisation, Circular No. 37 of 1988*, Government Printer, Lusaka.

GRZ (1989), *New Economic Recovery Programme. Fourth National Development Plan 1989-93*, National Commission For Development Planning, Lusaka, Zambia.

GRZ (1990), *1990 Census of Population. Housing and Agriculture - Preliminary Report*, C.S.O., Lusaka.

GRZ (1991a), *Policies. Programmes and Actions for Restructuring the Economy of Zambia: Statement by the Government of Zambia to the Donor/Creditor Community*, Paris, 11 December 1991.

GRZ (1991b), *Social Dimension of Adjustment. Priority Survey I. 1991*, C.S.O., Lusaka.

GRZ (1993), *Social Dimensions of Adjustment: Priority Survey II, Tabulation Report*, C.S.O., Lusaka.

GRZ (1994), *Social Dimensions of Adjustment, Priority Survey II, 1993, Tabulation Report*, C.S.O., Lusaka.

GRZ (1995a), *Internal Migration and Urbanisation: Aspects of 1990 Census of Population, Housing and Agriculture*, C.S.O., Population and Demographic Division, Lusaka.

GRZ (1995b), *Draft Housing Policy*, Ministry of Local Government and Housing, Lusaka.

GRZ (1995c), *Household Budget Survey 1993/94*, C.S.O., Lusaka.

Gulhati, R. (1991), 'Impasse in Zambia', *Public Administration and Development*, Vol. 11, pp. 239-44.

GVD (1996), *Government Valuation Department Quarterly Bulletin*, March/June, Research Development Unit, Lusaka.

Haan, H.C. (1982), *Some Characteristics of Informal Sector Business in Lusaka and Kitwe*, I.L.O., Zambia.

Haig, R. (1926), 'Toward an Understanding of the Metropolis', *Quarterly Journal of Economics*, Vol. 40.

Hailey, Lord (1957), *An African Survey*, Oxford University Press, Oxford.

Halcrow Fox and Associates (1989), *Zambia Land Delivery System*. In association with Clyed Surveys Limited. GRZ, Ministry of Water, Lands and Natural Resources, Lusaka.

Hall, B. (ed.). (1964), *Tell me, Josephine*, Andre Deutsch Ltd, 105 Great Russell Street, London.

Hammar, T. (1990), *Economic Ground Rent in Zambia: Principles for a New System*, Swedsurvey.

Hampden-Turner, C. (1991), *Charting the Corporate Mind*, Basil Blackwell.

Hanna, W.J. and Hanna, J.L. (1981), *Urban Dynamics in Black Africa: An Interdisciplinary Approach*, 2nd edn., Aldine, New York.

Hansen, K.T. (1997), *Keeping House in Lusaka*, Columbia University Press, New York.

Hansungule, M., Feeney, P. and Palmer, R. (1998), *A Report on Land Tenure Insecurity on the Zambian Copperbelt*. Prepared for Oxfam GB in Zambia.

Hawkesworth, N.R. (ed.) (1974), *Local Government in Zambia*, Lusaka City Council, Lusaka.

Haynes, J. (1995), 'The Revenge of Society? Religious Responses to Political Disequilibrium in Africa', feature review, *Third World Quarterly*, Vol. 16(4), pp. 728-36.

Haywood, I. (1985), 'Khartoum: City Profile', *Cities*, Vol. 2(3), pp. 186-97.

Haywood, I. (1986), 'Popular Settlements in sub-Saharan Africa', *Third World Planning Review*, Vol. 8(4), pp. 315-34.

Haywood, I. (1995), 'Why Plan for Public Participation?', *Housing and Planning Review*, April and May, pp. 6-7.

Healey, P. and Barrett, S.M. (1990), 'Structure and Agency in Land and Property Development Processes: Some Ideas for Research', *Urban Studies*, Vol. 27(1), pp. 89-104.

Heath, F.M.N. (1953), 'The Growth of African Councils on the Copperbelt', *Journal of African Administration*, Vol. 5(3), pp. 123-32.

Heisler, H. (1971), 'The Creation of a Stabilised Urban Society - A Turning Point in the Development of Northern Rhodesia/Zambia' *African Affairs*, Vol. 70(279), pp. 125-45.

Heisler, H. (1974), *Urbanisation and the Government of Migration*, C. Hurst and Company, London.

Herzer, H. and Pirez, P. (1991), 'Municipal Government and Popular Participation in Latin America', *Environment and Urbanization*, Vol. 3(1), pp. 79-95.

Heywood, A. (1994), *Political Ideas and Concepts - An Introduction*, Macmillan, London.

Holy Bible, New International Version.

Home, R. (1993), 'Barrack Camps for Unwanted People: a Neglected Planning Tradition', *Planning History*, Vol. 15(1), pp. 14-20.

Home, R., Ouzounova, T. and Rizov, M. (1996), *Turning Fish Soup Into Fish: Towards New Land and Property Markets in Bulgaria*, RICS Research Paper Series, Vol. 1(9).

Home, R. (1997), *Of Planting and Planning - The Making of British Colonial Cities*, Spon/Chapman and Hall, London.

Home, R. (1998), 'Barracks and Hostels - A Heritage Conservation Case for Worker Housing in Natal', *Natalia*, Vol. 28, pp. 45-52.

Home, R. and Jackson, J. (1997), *Land Rights for Informal Settlements: Community Control and the Single Point Cadastre in South Africa*, Our Common Estate Series, RICS, London.

Howes, C.K. (1980), *Value Maps: Aspects of Land and Property Values*, Geo Abstracts Ltd., Norwich.

Hughes, J.A. (1990). *The Philosophy of Social Research*, 2nd edn., Longman, London.

Hurd, R.M. (1924), *Principles of City Land Values*, The Record and Guide, New York.

Jameson, F.W. (1945), *Report on African Housing in Urban and Rural Areas, and Other Matters Allied With Housing in Northern Rhodesia*, filed at ZNA: SEC 1/1519, Local Government.

Kabisa, M. (1994), 'The Rating Act: 22 Stages to Preparing a Valuation Roll', paper presented to the Council Rating Workshop, 10-11 December 1994, Andrews Motel, Lusaka.

Kain, J. (1967), *Urban Form and the Costs of Urban Services*, Mimeograph. MIT-Harvard Joint Centre for Urban Studies, Cambridge, Massachusetts.

Kaluba, A. (1994), 'Kitwe: Zambia's Fast Growing City', *Times of Zambia*, 13 August, p. 7

Kanyeihamba, G.W. and McAuslan, J.P.W.B. (eds.) (1978), *Urban Legal Problems in Eastern Africa*, Scandinavian Institute of African Studies, Uppsala.

Kapumpa, M.S. (1994), *Submissions on the Amendments to the Rating Act. Cap. 484 of 1976*, 27 April 1994.

Kasongo, B.A. and Tipple, G.A. (1990), 'An Analysis of Policy Towards Squatters in Kitwe, Zambia', *Third World Planning Review*, Vol. 12(2), pp. 147-65.

Kaunda, K.D. (1967), *Humanism in Zambia and a Guide to its Implementation*, Part I, Government Printers, Lusaka, Zambia.

Kaunda, K.D. (1968), *Humanism in Zambia and a Guide to its Implementation*, Part I, Zambia Information Services, Lusaka.

Kaunda, K.D. (1970), 'The Exploitation of Man by Man'. Address to UNIP National Council, Mulungushi Hall, Government Printer, Lusaka.

Kaunda, K.D. (1974), *Humanism in Zambia and a Guide to its Implementation*, Part II, Division of National Guidance, Government Printers, Lusaka, Zambia.

Kaunda, K.D. (1975), Speech to the 6th UNIP National Council, 'The Watershed Speech', Zambia Information Services, 30 June 1975, Lusaka.

Kaunda, M. (no date), *Some Critical Issues in Land Policy Formulation in Zambia*, Copperbelt University, Kitwe, Zambia.

Kaunda, M., Kangwa, J. and Mutale, E. (1994), *Report of the Land Bill*, seminar held at Pamodzi Hotel, 23 September 1994, Copperbelt University, Kitwe, Zambia.

Kay, G. (1967), *A Social Geography of Zambia*, University of London Press, London.

KCC (1967), *The Kitwe Development Plan Written Document Modification No. 21*, 31 January 1967, Kitwe Planning Authority, Kitwe City Council, Kitwe.

KCC (1973a), *Official Opening of Water Works and Sewage Extension*, September 1973, Kitwe City Council, Kitwe.

KCC (1973b), *Policy on Squatter Settlements*, report of the City Engineer to a meeting of the Project Management Team, 9 January 1973, Kitwe City Council, Kitwe.

KCC (1976), *Low-cost Housing in Kitwe: Its Characteristics and Implications on Housing Policy*, Kitwe City Council, Kitwe.

KCC (1991a), *Letter from KCC's Financial Secretary to the General Manager of Nkana Division*, 31 December 1991.

KCC (1991b), *Annual Report for 1991*, Kitwe City Council, Kitwe.

KCC (1993/94), *Annual Report for the Years 1993 and 1994*, Kitwe City Council, Kitwe.

KCC (1994a), *Rating of ZCCM Properties - Amendments to the Rating Act Cap. 484.* Letter sent to the Committee for the Proposed Amendments to the Rating Act Cap. 484, by Kitwe City Council's Town Clerk, 27 April 1994.

KCC (1994b), *Report on the Water Supply and Sewerage Services*, June 1994, Kitwe City Council, Kitwe.

KCC (1995a), *Full Council Minutes for March 1995*, Kitwe City Council, Kitwe.

KCC (1995b), *Full Council Minutes for May 1995*, Kitwe City Council, Kitwe.

KCC (1995c), *Full Council Minutes for July 1995*, Kitwe City Council, Kitwe.

KCC (1995d), *Kitwe City Council Annual Report for 1995.* Kitwe City Council, Kitwe.

KCC (1997), *Full Council Minutes for February 1997*, Kitwe City Council, Kitwe.

KDC (1981), *Annual Report of the District Secretariat for the Year 1981*, Kitwe District Council, Kitwe.

KDC (1983), *Annual Report of the District Secretariat for the Year 1983*, Kitwe District Council, Kitwe.

KDC (1985), *Annual Report of the District Secretariat for the Year 1985*, Kitwe District Council, Kitwe.

Keare, D. and Parris, S. (1982), *Evaluation of Shelter Programmes for the Urban Poor - Principal Findings*, World Bank Staff Working Paper No. 547, Washington.

Kennedy, P. (1993), *Preparing for the 21st Century*, Harper Collins, London.

Ketz, R. (1992), 'The Exploitation of Copper in Zambia: Development and Environmental Consequences', *ITC Journal*, Vol. 2, pp. 130-33.

Kimani, S.M. (1972a), 'The Structure of Land Ownership in Nairobi', *Canadian Journal of African Studies*, Vol. 6(3), pp. 379-402.

Kimani, S.M. (1972b), 'Spatial Structure of Land Values in Nairobi, Kenya', *Economic and Social Geography*, Vol. 63, pp. 105-14.

King, A.D. (1976), *Colonial Urban Development: Culture, Social Power and Environment*, Routledge and Kegan Paul, London.

King, A.D. (1985), 'Colonial Cities: Global Pivots of Change', in R. Ross and G.J. Telkamp (eds.), *Colonial Cities: Essays on Urbanism in a Colonial Context*, Martin Nijhoff Publishers, Dordrecht, pp. 7-32.

King, A.D. (1990a), *Urbanism, Colonialism and the World Economy*, Routledge, London.

King, A.D. (1990b), *Global Cities. Post-Imperialism and the Internationalisation of London*, Routledge, London.

Kironde, J.M.L. (1992), 'Received Concepts and Theories in African Urbanisation and Management Strategies', *Urban Studies*, Vol. 29(8), pp. 1277-92.

Kirwan, R.M. (1989), 'Finance for Urban Public Infrastructure', *Urban Studies*, Vol. 26(3), pp. 285-300.

Kitwe Town Crier (1965a), 'Budget Speech: 1965 to See the Greatest Spending Yet', *Kitwe Town Crier*, Vol. 3, January, pp. 3-6. A monthly magazine of municipal news and views published by the Municipal Council of Kitwe, Kitwe, Zambia.

Kitwe Town Crier (1965b), '1966 Kitwe City Council Budget Speech', *Kitwe Town Crier*, November/December, pp. 3-5. A monthly magazine of municipal news and views published by the Municipal Council of Kitwe, Kitwe, Zambia.

Kitwe Water Supply Project (1991), *Kitwe Water Supply Project*, Vols. I-III, Kitwe, Zambia.

Knauder, S. (1982), *Shacks and Mansions: An Analysis of the Integrated Housing Policy in Zambia*, Mission Press, Ndola, Zambia.

Ladd, H.F. (1992), 'Population Growth, Density and the Costs of Providing Public Services, *Urban Studies*, Vol. 29(2), pp. 273-95.

Lacroux, S. (1997), 'Access to Land and Security of Tenure as Conditions for Sustainable Urban Development', UNCHS(Habitat), *Habitat Debate*, Vol. 3(2), June, pp. 11-12.

Lee-Smith, D. and Stren, R.E. (1991), 'New Perspectives on African Urban Management', *Environment and Urbanization*, Vol. 3(1), pp. 23-36.

Lungu, G.F. (1986), 'Mission impossible: Integrating Central and Local Administration in Zambia', *Planning and Administration*, Vol. 13(1), pp. 52-57.

Mabongunje, A.L. (1992), *Perspectives on Urban Land and Urban Management Policies in Sub-Saharan Africa*, World Bank Technical Paper No. 196, Africa Technical Department Series.

Majchrzak, A. (1984), *Methods for Policy Research*, Sage, London.

Makasa, K. (1981), *Zambia's March to Political Freedom*, Heinemann Educational Books, East Africa Ltd., Nairobi.

Malik, J.R., Branston, J. and Barraclough, J.H. (1974), 'The History and Finance of Local Government', in N.R. Hawkesworth (ed.), *Local Government in Zambia*, Lusaka City Council, Lusaka, pp. 13-32.

Marquard, L. (1933), 'The Problem of Government', in J.M. Davis (ed.), *Modem Industry and the African*, Macmillan, London, pp. 227-75.

Marquardt, M. (1995), 'Land Policy Issues in Uganda', *Land Tenure Centre Newsletter*, No. 73, University of Wisconsin, Madison.

Marshall, A. (1930), *Principles of Economics*, Macmillan, London.

Mattingly, M. (1993), 'Urban Management Intervention in Land Markets', in N. Devas and C. Rakodi (eds.), *Managing Fast Growing Cities*, Longman, pp. 102-31.

Mbao, M.L. (1987), *Law and Urbanisation in Zambia: A study of the Constitutional and Legal Framework of Urban Local Government 1890 to the Present*, Ph.D. thesis, University of Cambridge, Cambridge.

McAuslan, P. (1985), *Urban Land and Shelter for the Poor*, Earthscan, International Institute for Environment and Development.

McClain, W.T. (1978), 'Legal Aspects of Housing and Planning in Lusaka', in G.W. Kanyeihamba and J.P.W.B. McAuslan (eds.), *Urban Legal Problems in Eastern Africa*, Scandinavian Institute of African Studies, Uppsala, pp. 63-84.

Meebelo, H.S. (1986), *African Proletarians and Colonial Capitalism*, Kenneth Kaunda Foundation, Lusaka, Zambia.

Menezes, L. (1983), 'Public Land Disposal Methods', in UNCHS (Habitat), *United Nations Seminar of Experts on Land for Housing the Poor*, Tallberg, Sweden, 14-19 March 1983. In cooperation with the Ministry of the Interior, Finland and the Swedish Building Research Council, pp. 61-68.

Millward, R. (1988), 'Measured Sources of Inefficiency in the Performance of Private and Public Enterprises in LDCs', in P. Cook and C. Kirkpatrick (eds.), *Privatisation in Less Developed Countries*, Harvester Wheatsheaf, Brighton, pp. 143-61.

Ministry of Environment and Natural Resources (1992), *Zambia's National Report to the United Nations' Conference on Environment and Development*.

Ministry of Mines and Mineral Development (1992), *Investment Opportunities in the Mining Sector*, May 1992, Ministry of Mines and Mineral Development, Lusaka, Zambia.

Mitchell, J.C. (1956), *Power and Prestige Among Africans in Northern Rhodesia: An Experiment*, Rhodes-Livingstone Institute.

Mitchell, J.C. (no date), *A Note on the Urbanisation of Africans on the Copperbelt*, Institute for Economic and Social Research, University of Zambia, Lusaka.

MMD (1991), *Manifesto*, Campaign Committee, Movement for Multiparty Democracy Headquarters, Lusaka, Zambia.

Moser, C.A. and Kalton, G. (1971), *Survey Methods in Social Investigation*, 2nd edn., Heinemann Educational Books, London.

Mukonde, E.M. (1994a), 'Proposed Amendments to the Rating Act, 1976', paper presented to the Council Rating Workshop, 10-11 December 1994, Andrews Motel, Lusaka, Zambia. Published by the Ministry of Works and Supply, Government Valuation Department, Lusaka.

Mukonde, E.M. (1994b), 'The National Rating Programme and its Implementation', paper presented to the Council Rating Workshop, 10-11 December 1994, Andrews Motel, Lusaka, Zambia. Published by the Ministry of Works and Supply, Government Valuation Department, Lusaka.

Mukwena, R.M. (1992), 'Zambia's Local Administration Act, 1980: a Critical Appraisal of the Integration Objective', *Public Administration and Development*, Vol. 12, pp. 237-47.

Mulenga, S.P. (1981), 'National Development Programmes and Land Problems in Zambia', in CASLE (1981), *The Role of the Surveyor in Developing Countries with Particular Reference to Central and Southern Africa*, report of proceedings, Malawi, April, pp. 96-101.

Mulenga, S.P. (1983), 'Land Management Aspect of Property Legislation - The Zambian Experience', in CASLE (1983), *The Development of Africa Land Resources with Particular Reference to Central and Southern Africa*, report of proceedings, Zimbabwe, September, pp. 105-8.

Mulwanda, M. and Mutale, E. (1994), Never Mind the People, the Shanties Must Go: the Politics of Urban Land in Zambia', *Cities*, Vol. 11(5), pp. 303-11

Mumy, G.E. and Hanke, S.H. (1977), 'Optimal Departures from Marginal Cost Prices for Local Public Services', in P.B. Downing (ed.), *Local Service Pricing Policies and Their Effect on Urban Spatial Structure*, University of British Columbia Press, Vancouver, pp. 309-38.

Municipal Council of Kitwe (c.1959), *General Information Sheet*, file TC/59/76/1IMGB, filed at Zambia High Commission Library, London.

Municipal Council of Kitwe (1964), *Come, Let Us Build a City*, filed at the Zambia High Commission Library, London.

Municipality of Kitwe (1961), *Minute of His Worship the Mayor William McLachlan Comrie, for the Year Ended 31 March 1961*, filed at Zambia High Commission Library, London.

Municipality of Kitwe (1962), *Minute of His Worship the Mayor William McLachan Comrie J.P., for the Year Ended 31 March 1962*, filed at Zambia High Commission Library, London.

Mushota, R.K.K. (1993), 'The Movement for Multiparty Democracy and its Manifesto Provisions on Land Law and Policy Reforms for the Zambian Third Republic Government', paper presented to the National Conference on Land Policy and Legal Reform, July 1993, Lusaka.

Mutale, E. (1993), *Managing Rapid Urban Growth: With Particular Reference to Lusaka. Zambia*, unpublished M.Sc. thesis, University of East London, London.

Mutale, E. (1996), 'A Biblical View of Land Policy', *South African Journal of Surveying and Mapping*, Vol. 23, Part 6(142), December, pp. 325-32.

Mutale, E. (1999), *The Urban Development of Nkana-Kitwe, Zambia: Structural Conflict in the Management of Land and Services*, unpublished PhD thesis, University of East London, London.

Mvunga, M.P. (1982), *Land Law and Policy in Zambia*, University of Zambia Institute for African Studies, Zambian Papers No. 17.

Mwanza, J.M. (1979), 'Rural-Urban Migration and Urban Employment in Zambia', in B. Turok (ed.), *Development in Zambia*, Zed Press, London, pp. 26-36.

National Mirror (1991), 'Evicted Kanyama Residents Cry out "Our Marriages Are on the Rocks"', *National Mirror*, 16 December.

Netzer, D. (1988), 'Exactions in the Public Finance Context', in R. Alterman (1988), *Private Supply of Public Services*, New York University Press, New York, pp. 35-50.

Ndulo, M. (1987), *Mining Rights in Zambia*, Kenneth Kaunda Foundation, Lusaka, Zambia.

NHA (1974), *Questionnaire for National Urbanisation Strategy for Zambia*, National Housing Authority, Lusaka.

NIPA (1981), *Decentralisation: A Guide to Integrated Local Administration*, National Institute of Public Administration, Lusaka.

Northern Rhodesia (1917), *Mines Health and Sanitation Regulations*, Government Notice No. 17, Lusaka, Zambia.

Northern Rhodesia (1935), *Report of the Commission Appointed to Enquire Into the Disturbances in the Copperbelt of Northern Rhodesia*.

Northern Rhodesia (1944), *Report of the Commission on Administration and Finances of Native Locations in Urban Areas*, Government Printers, Lusaka.

Northern Rhodesia (1950), *Local Government and African Housing Annual Report for the Year 1949*, Government Printer, Lusaka, filed at ZNA: SEC 1/1519.

Northern Rhodesia (1957a), *Report of the Committee Appointed to Examine and Recommend Ways and Means by Which Africans Resident in Municipal and Township Areas Should be Enabled to Take an Appropriate Part in the Administration of Those Areas*, Government Printer, Lusaka.

Northern Rhodesia (1957b), *The Northern Rhodesia African Housing Board, First Annual Report*, filed at ZNA: SEC 1/1519, 'Local Government'.

Northern Rhodesia (1957c), *Report of the Committee on the Tenure of Urban Land in Northern Rhodesia*, Government Printer, Lusaka.

Northern Rhodesia (1960), *African Housing Board Annual Report of 1960*, Government Printer, Lusaka, filed at ZNA: SEC 1/1568.

Nyirenda, W.B. (1994), 'Planning and Controls in Local Authorities', in N.C. Deassis and S.M. Yikona (eds.), *The Quest for an Enabling Environment for Development in Zambia*, Mission Press, Ndola, Zambia, pp. 73-82.

O'Connor, A. (1983), *The African City*, Hutchinson University Library for Africa, London.

Oke, E.A. (1974), 'The Problems of Land Ownership in Zambia', paper presented for the final examination of the RICS.

Okpala, D.C.I. (1987), 'Received Concepts and Theories in African Urbanisation and Urban Management Strategies: A Critique', *Urban Studies*, Vol. 24(2), pp. 137-50.

Okpala, D.C.I. (1997), 'Of Planting and Planning: The Making of British Colonial Cities, Book Review', *Urban Studies*, Vol. 34(10), pp. 1744-47.

Ollowa, P.E. (1979), *Participatory Democracy in Zambia: The Political Economy of National Development*, Arthur H. Stockwell, Ilfracombe.

Open University (1980), 'Fundamentals of Human Geography, Section II, Spatial Analysis', *Social Sciences: A Second Level Course*, Units 9-21.

Oppenheim, N. (1980), *Applied Models in Urban and Regional Analysis*, Prentice-Hall, Englewood Cliffs, N.J.

Ottensmann, J.R. (1985), *Basic Microcomputer Programs for Urban Analysis and Planning*, Chapman and Hall, London.

Pakeni, C. (1973), *Social Action in Kitwe*, Mindolo Youth Leaders' Seminar, June, Mindolo Ecumenical Centre, Kitwe.

Palmer, R. (1998), "Back to the Land' at last? Privatisation of the Mines and Land Tenure Insecurity on the Zambian Copperbelt', paper presented to the African Studies Association of the UK Biennial Conference: Comparisons and Transitions, SOAS, University of London, 14-16 September.

Parkin, M. and King, D. (1992), *Economics*, Addison-Wesley, Wokingham.

Pasteur, D. (1978), 'Financial Resources for Urban Services and Development in Lusaka - A Case Study', *Occasional Papers in Development Administration*, No. 3, University of Birmingham, Birmingham.

Payne, G.K. (ed.) (1984), *Low-income Housing in the Developing World - the Role of Sites and Services and Settlement Upgrading*, John Wiley, Chichester.

Payne, G. (1989), *Informal Housing and Land Subdivisions in Third World Cities: A Review of the Literature*, CENDEP, Oxford.

Phillipson, D.W. (1977), *The Later Pre-history of East and South Africa*, Heinemann, London.

Platteau, J. (1992), *Land Reform and Structural Adjustment in sub-Saharan Africa: Controversies and Guidelines*, FAO, Economic and Social Development Paper 107.

Pollit, C. (1988), 'Bringing Consumers into Performance Measurement: Concepts, Consequences and Constraints', *Policy and Politics*, Vol. 16(2), pp. 77-87.

Prain, R.L. (1956), 'The Stabilisation of Labour in the Rhodesian Copperbelt', *African Affairs*, Vol. 55, pp. 305-12.

PRO: CO/795/36295.

PRO: CO/795/50/36295.

PRO: CO/795/62/5587.

PRO: CO/795/76/25628.

PRO: CO/795/76/45077.

PRO: CO/795/133/45384.

PRO: CO/795/68/25528.

PRO: CO/795/104/45260.

PRO: CO/795/110/45228.

Quick, S.A. (1975), *Bureaucracy and Rural Socialisation: The Zambian Experience*, Ph.D. thesis, Stanford University.

Rabinovitz, F.F. (1973), 'The Study of Urban Politics and the Politics of Urban Studies', in R. Wulff, R. (ed.), *Research Traditions in the Comparative Study of Urbanisation*, Comparative Urbanisation Studies, School of Architecture and Urban Planning, University of California, Los Angeles, pp. 83-100.

Rakodi, C. (no date), *Urban Land Policy in Zimbabwe*, Department of City and Regional Planning, University of Wales College of Cardiff.

Rakodi, C. (1986), 'Colonial Urban Policy and Planning in Northern Rhodesia and its Legacy', *Third World Planning Review*, Vol. 8(3), pp. 193-217.

Rakodi, C. (1987), 'Urban Plan Preparation in Lusaka', *Habitat International*, Vol. 11(4), pp. 95-111.

Rakodi, C. (1988a), 'Urban Agriculture: Research Questions and Zambian Evidence', *Journal of Modern African Studies*, Vol. 26(3), pp. 494-515.

Rakodi, C. (1988b), 'The Local State and Urban Local Government in Zambia', *Public Administration and Development*, Vol. 8, pp. 27-46.

Rakodi, C. (1991), 'Developing Institutional Capacity to Meet the Housing Needs of the Poor - Experience in Kenya, Tanzania and Zambia', *Cities*, Vol. 8(3), pp. 228-243.

Rapoport, A. and Hardie, G. (1991), 'Cultural Change Analysis - Core Concepts of Housing for the Tswana', in: A.G. Tipple and K.G. Willis (eds.), *Housing the Poor in the Developing World - Methods of Analysis, Case Studies and Policy*, Routledge, London, pp. 35-61.

Rawls, J. (1971), *A Theory of Justice*, Harvard University Press, Cambridge, Massachusetts.

Retrospect (1962), *A Report by His Worship (Councillor W.M. Comrie, J.P.) on the Work of the Municipal Council of Kitwe During the 1962 Civic Year*, filed at Zambia High Commission Library, London.

Retrospect (1963), *An Account of the Work of the Municipal Council of Kitwe During the Civic Year 1963/64, by Councillor C.R. Wilkins, Mayor of Kitwe*, filed at Zambia High Commission Library, London.

Roan Consolidated Mines Ltd. (1978), *Zambia's Mining Industry - The First 50 years*, Open File report, Lusaka.

Robinson, E.A.G. (1933), 'The Economic Problem', in J.M. Davis (ed.), *Modem Industry and the African*, Macmillan, London, pp. 131-224.

Rondinelli, D.A. (1981), "Government Decentralisation in Comparative Perspective: Theory and Practice in Developing Countries', International Review of Administrative Sciences, Vol. XI. VII(2), pp. 133-45.

Ross, R. and Telkamp, G.J. (eds.) (1985), *Colonial Cities: Essays on Urbanism in a Colonial Context*, Martin Nijhoff Publishers, Dordrecht.

Roth, M. (ed.) (1995), *Land Tenure. Land Markets and Institutional Transformation in Zambia*, research report, Land Tenure Centre, University of Wisconsin-Madison.

Roth, M., Khan, A.M. and Zulu, M.C. (1995), 'Legal Framework and Administration of Land Policy in Zambia', in M. Roth (ed.), *Land Tenure. Land Markets and Institutional Transformation in Zambia*, Land Tenure Centre, University of Wisconsin-Madison, pp. 1-46.

Sandbrook, R. (1988), 'Patrimonialism and the Failing of Parastatals: Africa in Comparative Perspective', in P. Cook and C. Kirkpatrick (eds.), *Privatisation in Less Developed Countries*, Harvester Wheatsheaf, Brighton, pp. 162-79.

Sanyal, H. (1981), *Urban Agriculture: a Strategy for Survival in Zambia*, Ph.D. thesis, University of California, Los Angeles.

Sazanami, H. (1984), 'Keynote Address', in G.S. Cheema (ed.), *Managing Urban Development: Services For the Poor*, United Nations Centre for Regional Development, Nagoya, pp. 3-9.

Schlyter, A. (1984), *Upgrading Reconsidered - The George Studies in Retrospect*, The National Swedish Institute for Building Research, Gavle, Sweden.

Schlyter, A. (1988), *Women Householders and Housing Strategies: The Case of George, Zambia*, Research Report, The National Swedish Institute for Building Research, Gavle, Sweden.

Schlyter, A. and Schlyter, T. (1979), *George - The Development of a Squatter Settlement in Lusaka, Zambia*, The National Swedish Institute for Building Research, Gavle, Sweden.

Schwandt, T.A. (1994), 'Constructivist, Interpretivist Approaches to Human Inquiry', in N.K. Denzin and Y.S. Lincoln (eds.), *Handbook of Qualitative Research*, Sage Publications, London, pp. 118-37.

Seymour, T. (1975), 'Squatter Settlement and Class Relations in Zambia', *Review of Political Economy*, No. 3, pp. 71-77.

Sheppard, D. (1983), *Bias to the Poor*, Hodder and Stoughton, London.

Shevky, E. and Bell, W. (1955), *Social Area Analysis: Theory, Illustration, Application and Computational Procedures*, Stanford University Press, California.

Sills, D.L. (ed.) (1968), *International Encyclopaedia of Social Sciences*, Vol. 3, Macmillan, New York.

Simmance, A.J.F. (1974), 'The Structure of Local Government in Zambia', in N.R. Hawkesworth (ed.), *Local Government in Zambia*, Lusaka City Council, Lusaka, pp. 4-12.

Simons, H.J. (1979), 'Zambia's Urban Situation', in B. Turok (ed.), *Development in Zambia*, Zed Press, London, pp. 1-25.

Sirken, I.A. (1982), 'Introduction to Urban Taxation; Financing Urban Development: Towards More Resources and Better Management', *The Urban Edge*, Vol. 6(6), p. 1.

Slinn, P. (1972), 'The Legacy of the British South African Company: The Historical Background', in M. Bostock and C. Harvey (eds.), *Economic Independence and Zambian Copper*, Praeger, New York, pp. 23-52.

Smith, A. (1937), *Wealth of Nations*, Modem Library, New York.

Snyder, T. and Stegman, M. (1987), *Paying for Growth*, Urban Land Institute, Washington DC.

Stren, R. and White, R. (eds.) (1989), *African Cities in Crisis: Managing Rapid Urban Growth*, Westview Press, Boulder.

Stren, R.E. (1991), 'Old Wine in New Bottles? An Overview of Africa's Urban Problems and the 'Urban Management' Approach to Dealing With Them', *Environment and Urbanisation*, Vol. 3(l), pp. 9-22.

Stren, R.E. (1992), 'African Urban Research Since the Late 1980s: Responses to Poverty and Urban Growth', *Urban Studies*, Vol. 29(3/4), pp. 533-55.

Sunday Times of Zambia (1997), 'Chaos Reigns Over Plots Allocation', *Sunday Times of Zambia*, 14 September, p.1.

Swanson, M.W. (1976), 'The Durban System: Roots of Urban Apartheid in Colonial Natal', *African Studies*, Vol. 35(3/4), pp. 159-76.

The Weekly Post (1992a), 'What are Market Forces', *The Weekly Post*, 13-19 March.

The Weekly Post (1992b), 'Labour Party Supports Strike', *The Weekly Post*, 2-8 October.

The Post (1994), 'Lusaka Council to Stop Illegal Plot Allocation', *The Post*, 14 October.

The Post (1998a), 'Urban Poverty on the Increase', *The Post*, 6 April 1998.

The Post (1998b), Scott Questions Tembo's Thinking', *The Post*, 22 December.

Thirkell, A. (1989), *Urban Land World Bank Policy: Redressing the Balance Between the State, the Market and the Poor*, Development Planning Unit, Working Paper No. 38.

Times of Zambia, (1990), 'Rent Payers Team Up', *Times of Zambia*, 30 December.

Times of Zambia (1994a), 'Property owners breath fire', *Times of Zambia*, 28 February.

Times of Zambia (1994b), 'Kitwe Owed K4bn', *Times of Zambia*, , 13 August.

Times of Zambia (1994c), 'Lusaka Plots Row Erupts'. *Times of Zambia*, 15 August, p. 3.

Times of Zambia (1994d), 'Council Water Squad Chased', *Times of Zambia*, 16 August.

Times of Zambia (1994e), '5 Councils' Ratings Thrown Out', *Times of Zambia*, 4 October, p. 3.

Times of Zambia (1994f), 'Pay up or Else, Warns Council', *Times of Zambia*, 11 October.

Times of Zambia (1996), 'Chongwe Council Rates Too High', *Times of Zambia*, 16 October.

Times of Zambia (1997a), 'Loan Seekers Flood ZNBS', *Times of Zambia*, 9 July.

Times of Zambia (1997b), 'Opinion Column', *Times of Zambia*, 6 August.

Times of Zambia (1997c), 'Opinion Column', *Times of Zambia*, 20 October.

Times of Zambia (1997d), 'Chiluba Extends House Sales Deadline', *Times of Zambia*, 11 December.

Times of Zambia (1998a), 'Give us More Time, Mr. President', letter to the editor by worried civil servant of Lusaka, *Times of Zambia*, 25 February.

Times of Zambia (1998b), 'Lusaka Woes Stay Unless...', *Times of Zambia*, 28 February.

Times of Zambia (1998c), 'Unions Hail State Move', *Times of Zambia*, 1 September.

Tipple, A.G. (1976a), 'The Low-cost Housing Market in Kitwe, Zambia', *Ekistics*, 244, pp. 148-52.

Tipple, A.G. (1976b), 'Self-Help Housing Policies in a Zambian Mining Town', *Urban Studies*, Vol. 13, pp. 167-69.

Tipple, A.G. (1978), 'Low-cost Housing Policies in the Copperbelt Towns of Northern Rhodesia/Zambia: An Historical Perspective', in A.V. Akeroyd and C.R. Hill (eds.), *Southern Africa Research in Progress*, Collected Papers 4, Centre for Southern Africa Studies, University of York, pp. 149-63.

Tipple, A.G. (1981), Colonial Housing Policy and the 'African Towns' of the Copperbelt: The Beginnings of Self Help', *African Urban Studies*, 11, pp. 65-85.

Tipple, A.G. and Willis, K.G. (eds.) (1991), *Housing the Poor in the Developing World - Methods of Analysis, Case Studies and Policy*, Routledge, London.

Todd, D. and Sinjwala, E. (1985), *Evaluation of DANIDA/UNCHS Training Programme for Community Participation in Improving Human Settlements in Zambia*, Institute for African Studies, University of Zambia, Lusaka.

Tordoff, W. (ed.) (1980), *Administration in Zambia*. Manchester University Press, Manchester.

Town and Country Planning (1979), *Human Settlement Policies in Zambia. Post Habitat 1976*, Department of Town and Country Planning Research Unit, September, Lusaka, Zambia.

Trompenaars, F. (1993), *Riding the Waves of Culture - Understanding Diversity in Business*, Economics Books, London.

Turok, B. (ed.) (1979), *Development in Zambia*, Zed Press, London.

Turok, B. (1989), *Mixed Economy in Focus: Zambia*, Institute for African Alternatives, London.

UN (1966), *Progress in Land Reform - Fourth Report*, United Nations, New York.

UN (1967), *Methods of Estimating Housing Needs*, Studies in Methods, Series F, No. 12, United Nations, New York.

UN (1973), *Urban Land Policies and Land-Use Control Measures*, Vol. 1, Africa, United Nations, New York.

UNCHS (Habitat) (1974), *Land for Human Settlements. Review and Analysis of the Present Situation. Recommendations for National and International Action*, Habitat, Nairobi, Kenya.

UNCHS (Habitat) (1983), *United Nations Seminar of Experts on Land for Housing the Poor*, Tallberg, Sweden, 14-19 March 1983. In cooperation with the Ministry of the Interior, Finland and the Swedish Building Research Council.

UNCHS (Habitat) (1990), *Guidelines for the Improvement of Land-Registration and Land-Information Systems in Developing Countries (With Special Reference to English-speaking Countries in Eastern, Central and Southern Africa)*, Nairobi, United Nations Center for Human Settlements.

UNCHS (Habitat) (1996a), *Habitat Agenda and Istanbul Declaration*, United Nations Department of Public Information, New York.

UNCHS (Habitat) (1996b), *New Delhi Declaration. Global Conference on Access to Land and Security of Tenure as a Condition For Sustainable Shelter and Urban*

Development, New Delhi, India, 17-19 January 1996, preparation document for Habitat 11, Istanbul, June, 1996.

UNCHS (Habitat) (1997), 'Environmental Studies', *Habitat Debate*, Vol. 3(2), June.

Van den Berg, L.M. (1974), *Kitwe: The Growth of a Mining Town*, Zambia Geographical Association, Handbook Series: No. 3.

Van den Berg, L.M. (1984), *Anticipating Urban Growth in Africa - Land Use and Land Values in the Rurban Fringe of Lusaka. Zambia*, Zambia Geographical Association, Occasional Study No. 13.

Velasquez, C. (1991), 'Local Governments in Intermediate Sized Cities in Colombia: Municipal Governments for Whom?', *Environment and Urbanisation*, Vol. 3(l), pp. 109-20.

Water Sector Development Group (1993), Website http://www.Zamnet.zm/zamnet/water/wsdg/wsdg.html

Wheaton, W.L.C. and Schussheim, M.J. (1955), *The Costs of Municipal Services in Residential Areas*, US Department of Commerce, Washington, DC.

Williams, G. (ed.) (1986), *Lusaka and Its Environs: A Geographical Study of a Planned Capital in Tropical Africa*, Zambia Geographical Association Handbook Series No. 9.

Windsor, D. (1979), 'A Critique of the Costs of Sprawl', *Journal of American Planning Association*, 45, pp. 279-92.

World Bank (2002), *African Development Indicators 2002*, World Bank, Washington DC.

Wright, S. and Nelson, N. (eds.) (1995), *Power and Participatory Development: Theory and Practice*, Intermediate Technology, London.

Wrong, M. (1997), 'Zambia - 'Crown Jewel' Earmarked for Foreign Buyers', *London Financial Times Survey*, 4 March.

Wulff, R. (ed.) (1973), *Research Traditions in the Comparative Study of Urbanization*, Comparative Urbanization Studies, School of Architecture and Urban Planning, University of California, Los Angeles.

Zambia Daily Mail (1977), *Zambia Daily Mail*, 29 August.

Zambia Daily Mail (1994), 'Comment', *Zambia Daily Mail*, 9 March.

Zambia Daily Mail (1996a), 'Town Clerk Simwinga Advises Residents', *Zambia Daily Mail*, 3 April.

Zambia Daily Mail (1996b), 'House Owners Demand Refund', *Zambia Daily Mail*, 17 June 1996.

Zambia Daily Mail (1997), 'Illegal Settlers Granted Injunction', *Zambia Daily Mail*, 20 October.

Zambia Daily Mail (1998), '20 Kitwe Houses Collapse', *Zambia Daily Mail*, 18 March.

Zambia Today (1996), 'Policy on Decentralization Coming', *Zambia Today*, electronic news from Zambia, 18 July.

Zambia Today (1997a), 'CBU Chemical School in Trouble'. *Zambia Today*, electronic news from Zambia, 10 February, Website http://www.zamnet.zm/zamnet/zanalzanabase/970210.cbu.html

Zambia Today (1997b), 'Lusaka City Rate Levy goes down', *Zambia Today*, electronic news from Zambia, 26 June, Website http://www.zamnet.zm/zamnet/zanalzanabase/970626.levy

ZCCM (1994), *Rateable Properties - ZCCM Nkana Division*, letter from the Head of Finance, Nkana Division, to the Director of Finance, KCC, 8 June 1994.

ZCCM (1997), *Nkana Mining Licence Area - ML3, Environmental Impact Statement, Appendix C, Socio-Economic Issues*, Zambia Consolidated Copper Mines, Kitwe, Zambia.

ZNA: RC/1427, *The Establishment of a Public Township at Nkana*.

ZNA: SEC 1/1320, *Stabilization of African Labour.*

ZNA: SEC 1/1519, *Local Government.*

ZNA: SEC 1/1528, *Kitwe - Establishment of Public Township. File No. LG/KIT/5 '.*

ZNA: SEC 1/1535, Vols. I-II, *Nkana Mine Township.*

ZNA: SEC 1/1568, *Location Superintendents' Conference*, February - March 1950, Lusaka.

ZNA: SEC 1/1568, *Location Superintendents' Conference*, February 1951, Ndola.

ZNBC (1994), *Points of View*, television programme, presented by Alexander Miti from Kitwe Studios, 9 September 1994 at 20.30 hours.

Appendix Table 1 Mains water consumption in Nkana-Kitwe, 1974

Area	Est. no. of consumers or households	Est. total pop'ln[a] first and sub-households	Servants' quarters	Daily total peak consumption (cmd)	Daily avr. consumer / household (cmd)
Industrial area	600	N/A		9,565	15.9
City centre					
CBD/Martindale	650	2,000[b]	12,000[c]	4,140	6.4
Low-cost					
Bulangililo	955	6,102		1,868	2.0
Chimwemwe	5,800	37,062		13,825	2.4
Kamitondo/Buchi	2,250	14,378		3,886	1.7
Kwacha/K. East	3,220	20,576		6,591	2.0
Luangwa	1,050	6,710		1,494	1.4
Ndeke Township	2,900	18,531		5,978	2.1
Twatasha R/course	161	1,029		306d	1.9
High-cost					
East of CPC+ CPC	305	1,949		3,288	10.8
Nkana East	1,200	7,668		5,380	4.5
Parklands	950	6,070		5,156	5.4
Riverside	1,000	6,390		5,306	5.3
Informal					
Itimpi (Informal)	875	5,591		1662[d]	1.9
Mulenga	226	1,444		429[d]	1.9
Mwaiseni	339	2,166		644[d]	1.9
Outer suburb					
Itimpi (Formal)	160	1,022		2,586	16.2
Sub-total Kitwe	*22,641*	*138,688*	*12,000*	*72,104*	*3.2*
Low-cost					
Chamboli	2,560	17,152[e]	}		
Mindolo	2,276	15,249[e]	}		
Mindolo North	526	3,524[e]	}		
Wusakile	4,410	29,547[e]	}		
Twibukishe	29	194	}		
Medium-cost			}	44,552[h]	4.0
Miseshi	556	4,170[f]	}		
Twibukishe	90	675[f]	}		
Wusakile East	125	938[f]	}		
High-cost			}		
Nkana West	504	3,220[g]	}		
Sub-total Nkana	11,076	7,4669			
Nkana-Kitwe	33,717	213,357	12,000	116,656	3.4

Sources: City of Kitwe Dev. Plan (1975-2000), Final Report p.197; City of Kitwe Dev. Plan (1975-2000), Survey of Existing Conditions Tables 3.10-3.12; NHA Questionnaire for National Urbanization Strategy.

Notes: a. estimated from an occupancy rate of 6.39 persons per dwelling (NHA, 1974, Table 2 item x)

   b. estimated from avr. 6.39 persons/dwelling from 313 residential stands

   c. from NHA, 1974, Table 2 item iii

   d. based on estimated daily consumption of 1.9 cubic metres per day/household from the observed (City of Kitwe Dev. Plan 1975-2000 Final Report) mean daily consumption of low-cost households in Kitwe

   e. based on 6.7 persons per dwelling (NHA, 1974, Table 1A item v)

   f. based on 7.5 per house (NHA, 1974, Table 2 item iv)

   g. based on 6.39 persons per dwelling (NHA, 1974, Table 2 item x)

   h. denotes remainder from Nkana's 63,645 cmd production after 30 per cent losses

Appendix Table 2 **Nkana-Kitwe estimated water demand in 1990 based on the observed 1974 consumption rates**

Area	Estimated number of consumers or households	Estimated total population[a] first and sub-households	Estimated daily total peak consumption (cmd)[b]	Estimated daily average per consumer or household (cmd)
Industrial area	600	N/A	9,540	15.9
City centre				
CBD/Martindale	1,063	5,496	6,803	6.4
Low-cost				
Buchi	1,186	8,437	2,016	1.7
Bulangililo	1,435	12,638	2,870	2.0
Chimwemwe	7,089	44,333	17,014	2.4
Kamitondo	1,125	7,181	1,912	1.7
Kawama	640	4,098	1,216	1.9
Kwacha	3,212	22,004	6,424	2.0
Luangwa	1,790	11,648	2,506	1.4
Ndeke Township	2,892	22,334	6,073	2.1
Ndeke Village	1,489	7,394	2,829	1.9
Twatasha R/course	2,300	14,287	4,370	1.9
High-cost				
Nkana East	1,390	12,411	6,255	4.5
Parklands	1,650	10,586	8,910	5.4
Riverside	1,650	10,611	8,745	5.3
Informal				
Ipusukilo	3,000	14,592	5,700	1.9
Itimpi (Informal)	800	See note c	1,520	1.9
Mulenga	1,605	7,979	3,050	1.9
Mwaiseni	480	2,584	912	1.9
Outer suburb				
Itimpi (Formal)	500	8,110[c]	8,100	16.2
Sub-total Kitwe	*35,896*	*226,723*	*106,765*	*3.0/consumer*
Low-cost				
Chamboli	2,433	20,676	}	
Mindolo	2,386	18,652	}	
Wusakile	3,919	23,525	}	
Medium-cost			}	
Miseshi/ChaChaCha	780	6,189	}45,437	4.0/consumer
Natwange	710	4,555	}	
Twibukishe/Plant	447	3,024	}	
High-cost			}	
Nkana West	639	4,550	}	
Sub-total Nkana	11,296	81,171	45,437	
Nkana-Kitwe	47,192	307,894	152,202	3.2/consumer

Sources: Author's own estimates based on observed 1974 consumption patterns.

Notes:

a. population figures from Kitwe Water Supply Project 1991, Table 3.3-1

b. daily total peak consumption is the product of the respected number of consumers and the corresponding 1974 rate of consumption from Appendix Table 1. Assumes a uniform elasticity in consumption across the consumer spectrum between 1974 and 1990

c. population in the outer suburb of Itimpi includes that of the Itimpi informal area

Appendix Table 3   Nkana-Kitwe estimated peak consumption in 1990 adjusted for network losses

Area	Estimated number of consumers or households	Estimated total population[a] first and sub-households	Estimated daily total peak consumption (cmd)[b]	Estimated daily average per consumer or household (cmd)
Industrial area	600	N/A	6,434	10.7
City centre				
CBD/Martindale	1,063	5,496	4,588	4.3
Low-cost				
Buchi	1,186	8,437	1,360	1.1
Bulangililo	1,435	12,638	1,936	1.3
Chimwemwe	7,089	44,333	11,475	1.6
Kamitondo	1,125	7,181	1,289	1.1
Kawama	640	4,098	820	1.3
Kwacha	3,212	22,004	4,333	1.3
Luangwa	1,790	11,648	1,690	0.9
Ndeke Township	2,892	22,334	4,096	1.4
Ndeke Village	1,489	7,394	1,908	1.3
Twatasha R/course	2,300	14,287	2,947	1.3
High-cost				
Nkana East	1,390	12,411	4,219	4.5
Parklands	1,650	10,586	6,009	3.6
Riverside	1,650	10,611	5,898	3.6
Informal				
Ipusukilo	3,000	14,592	3,844	1.3
Itimpi (Informal)	800}		1,025	1.3
Mulenga	1,605	7,979	2,057	1.3
Mwaiseni	480	2,584	615	1.3
Outer suburb				
Itimpi (Formal)	500}	8,110	5,463	10.9
Sub-total Kitwe	*35,896*	*226,723*	*72,006*	*2.0/consumer*
Low-cost				
Chamboli	2,433	20,676	}	
Mindolo	2,386	18,652	}	
Wusakile	3,919	23,525	}	
Medium-cost			}	
Miseshi/ChaChaCha	780	6,189	}30,645	2.7/consumer
Natwange	710	4,555	}	
Twibukishe/Plant	447	3,024	}	
High-cost			}	
Nkana West	639	4,550	}	
Sub-total Nkana	11,296	81,171	30,645	
Nkana-Kitwe	47,192	307,894	102,651	2.2/consumer

Sources: Author's own estimations based on Appendix Table 2.

Appendix Table 4   Estimated charges and potential revenue based on the 1990 estimated consumption and 1994 tariff

Area	No. of consumers	Daily total peak consumption (cmd)	Total Monthly consumption per customer (m³)	Monthly Tariff charge per customer (K)	Total tariff charge per month (million K)	Unit charge (K/m³) x no. of customers
Industrial area	600	6,434	322	91,250	55	283x600
City centre						
CBD/Martindale	1,063	4,588	129	40,050	42	310x1,063
Low-cost						
Buchi	1,186	1,360	34	2,000	2.4	59 x1,186
Bulangililo	1,435	1,936	40	12,750	18	319x1,435
Chimwemwe	7,089	11,475	48	14,750x89	15.3	307x89
				2,000x7,000		42x7,000
Kamitondo	1,125	1,289	34	2,000	2.2	59x1,125
Kawama	640	820	38	12,250x100	2.1	322x100
				2,000x340		53x340
				800x200		21x200
Kwacha	3,212	4,333	40	12,750x879	15.9	319x879
				2,000x2,333		50x2,333
Luangwa	1,790	1,690	28	2,000x540	2.1	71x540
				800x1,250		28x1,250
Ndeke Township	2,892	4,096	42	13,250	38	315x2,892
Ndeke Village	1,489	1,908	38	12,250	18	322x1,489
Twatasha R/course	2,300	2,947	38	12,250x100	3.4	322x100
				2,000x300		53x300
				800x1,900		21x1,900
High-cost						
Nkana East	1,390	4,219	91	25,500	36	280x1,390
Parklands	1,650	6,009	109	30,000	50	275x 1,650
Riverside	1,650	5,898	107	29,500	50	276x1,650
Informal						
Ipusukilo	3,000	3,844	38	2,000	6	53x3,000
Itimpi (Informal)	800	1,025	38	800	0.6	21x800
Mulenga	1,605	2,057	38	800	1.3	21x1,605
Mwaiseni	480	615	38	800	0.4	21x480
Outer suburb						
Itimpi (Formal)	500	5,463	328	84,750	42	258x500
Total	**35,896**	**72,006**			**408.7**	

Notes: Monthly tariff charges have been calculated using consumption levels from Appendix Table 3 and the 1994 tariff (Appendix Table 8). Unit charges reflect the price for 1m³ of consumption. For example, for Chimwemwe, there are two groups of consumers, one group of 89 households have meters and pay an estimated K14,750 per month. The other group of 7,000 have private stand-pipes and pay a fixed charge of K2,000 per month, but the average consumption for both is estimated at 48 cmd. Unit charges for the respective groups are (i) K14,750/48m³ = K307/m³ and (ii) K2,000/48m³ = K42/m³.

Appendix Table 5   Estimated all-meter charges based on the 1990 estimated consumption and 1994 tariff

Area	No. of consumers	Daily total Peak consumption (cmd)	Total monthly consumption per customer (m³)	Monthly tariff charge per customer (K)	Total tariff charge per month (million K)	Unit charge (K/m²) x no. of customers
Industrial area	600	6,434	322	91,250	55	283
City centre						
CBD/Martindale	1,063	4,588	129	40,050	42	310
Low-cost						
Buchi	1,186	1,360	34	11,250	13	331
Bulangililo	1,435	1,936	40	12,750	18	319
Chimwemwe	7,089	11,475	48	14,750	106	307
Kamitondo	1,125	1,289	34	11,250	12	331
Kawama	640	820	38	12,250	8	322
Kwacha	3,212	4,333	40	12,750	42	319
Luangwa	1,790	1,690	28	9,750	18	348
Ndeke Township	2,892	4,096	42	13,250	38	315
Ndeke Village	1,489	1,908	38	12,250	18	322
Twatasha R/course	2,300	2,947	38	12,250	28	322
High-cost						
Nkana East	1,390	4,219	91	25,500	36	280
Parklands	1,650	6,009	109	30,000	50	275
Riverside	1,650	5,898	107	29,500	50	276
Informal						
Ipusukilo	3,000	3,844	38	12,250	36	322
Itimpi (Informal)	800	1,025	38	12,250	10	322
Mulenga	1,605	2,057	38	12,250	19	322
Mwaiseni	480	615	38	12,250	6	322
Outer suburb						
Itimpi (Formal)	500	5,463	328	84,750	42	258
Total	**35,896**	**72,006**			**648**	

Source: Author's calculations based upon estimated consumption and 1994 Kitwe tariff.

Appendix Table 6 Comparison between current pricing policy and all-meter pricing

Area	Number of consumers	Current policy unit charge (K/m^3) x number of consumers	All-meter with factor unit charge (K/m^3)	Gainers	Losers
Industrial area	600	283	174	600	
City centre					
CBD/Martindale	1,063	310	191	1,063	
Low-cost					
Buchi	1,186	59	204		1,186
Bulangililo	1,435	319	196	1,435	
Chimwemwe	7,089	307x89	189	89	7,000
		42x7,000			
Kamitondo	1,125	59	204		1,125
Kawama	640	322x100	198	100	
		53x340			340
		21x200			200
Kwacha	3,212	319x879	196	879	
		50x2,333			2,333
Luangwa	1,790	71x540	214		540
		28x1,250			1,250
Ndeke Township	2,892	315	194	2,892	
Ndeke Village	1,489	322	198	1,489	
Twatasha	2,300	322x100	198	100	
R/course		53x300			300
		21x1,900			1,900
High-cost					
Nkana East	1,390	280	172	1,390	
Parklands	1,650	275	169	1,650	
Riverside	1,650	276	170	1,650	
Informal					
Ipusukilo	3,000	53	198		3,000
Itimpi (Informal)	800	21	198		800
Mulenga	1,605	21	198		1,605
Mwaiseni	480	21	198		480
Outer suburb					
Itimpi (Formal)	500	258	159	500	
Total	**35,896**			**13,837**	**22,059**

Sources: Based on author's calculations from Appendix Tables 4 and 5.

Note: To determine the number of gainers and losers: for example, in Chimwemwe, there are currently 89 households on a metered supply paying an average of $K307/m^3$. Under all-meter, these would be paying $K189/m^3$, hence there would be 89 gainers in Chimwemwe. Similarly, there are 7,000 households paying $K42/m^3$ who would have to pay $K189/m^3$, representing an increase in charges (or loss of income) of $K147/m^3$, hence 7,000 losers in Chimwemwe.

Appendix Table 7 Marginal cost pricing

Area	Dist. from reservoir (Km)	No. of consu- mers	Daily total peak consump'n (cmd)	Marginal distrib'n cost (000K)	Marginal prod'n cost (000K)	Total Marginal Cost (000K)	Unit charge (K/m³)
Industrial	1.0	600	6,434	161	804	965	150(1,608)
City centre							
CBD / Martindale	2.1	1,063	4,588	241	574	815	178(767)
Low-cost							
Buchi	0.7	1,186	1,360	24	170	194	143(164)
Bulangililo	2.8	1,435	1,936	136	242	378	195(263)
Chimwemwe	2.4	7,089	11,475	688	1,434	2,122	185(299)
Kamitondo	1.4	1,125	1,289	45	161	206	160(183)
Kawama	1.4	640	820	29	102	131	160(205)
Kwacha	2.8	3,212	4,333	303	542	845	195(263)
Luangwa	5.6	1,790	1,690	237	211	448	265(250)
Ndeke Township	1.8	2,892	4,096	184	512	696	170(170)
Ndeke Village	1.8	1,489	1,908	86	238	324	170(218)
Twatasha Racecourse	1.8	2,300	2,947	133	368	501	170(218)
High-cost							
Nkana East	1.4	1,390	4,219	148	527	675	160(486)
Parklands	1.4	1,650	6,009	210	751	961	160(582)
Riverside	2.4	1,650	5,898	354	737	1,091	185(661)
Informal							
Ipusukilo	2.4	3,000	3,844	231	480	711	185(237)
Itimpi (Informal)	2.1	800	1,025	54	128	182	178(228)
Mulenga	3.2	1,605	2,057	164	257	421	205(262)
Mwaiseni	3.2	480	615	49	77	126	205(262)
Outer suburb							
Itimpi (Formal)	1.4	500	5,463	191	683	874	160(1,748)
Total		**35,896**	**72,006**	**3,668**	**8,998**	**12,666**	
Average	**2.2**		2 cmd/ consumer	K51/m³ (K102/ consumer)	K125/m³ (K251/ consumer)	Total avr. cost = K176/m³	

Source: Author's calculations based upon estimated consumption, scaled distances and 1994 tariff.

Notes: Distribution costs are calculated using average distance from reservoir for the group of consumers. For example, for Chimwemwe, given that distribution cost is K25/cmd/Km, marginal distribution costs are (K25 x 11,475 cmd x 2.4Km) / (cmd x Km) = K688,500.

Figures enclosed within brackets in the last column of the table refer to the charge per consumer as opposed to the other one which refers to the charge per unit consumed. Relating charges to population offers an alternative approach to recovering marginal costs by using a flat rate without necessarily working out individual costs based on consumption. In terms of potential for cost recovery, the flat rate approach would lead to complaints from consumers who might regard their consumption as being lower than others.

Appendix Table 8   Kitwe water tariff – 1994

User	Water Tariff (K)	Sewerage Tariff (K)
Domestic:		
Tariff 32 and 36		5,000 per point
first 15 units	7,000	
next 10 units	2,000	
additional units	250	
stuck meters	15,000	
Industrial/commercial:		
Tariff 18, 34 and 38		15,000 per point
first 80 units	18,000	
next 80 units	350	
additional units	450	
stuck meters	15,000	
Domestic:		
Imperial system:		
Tariff 12, 13 and 16		5,000 per point
first 6 units	10,000	
next 4 units	600	
additional units	650	
stuck meters	15,000	
Industrial/commercial:		
Tariff 14 and 17		15,000 per point
first 20 units	22,000	
next 20 units	900	
additional units	1,300	
stuck meters	45,000	
Low-cost areas - townships:		
Fixed charge		
	2,000	2,000 per point

Appendix Table 9   Revised Kitwe water tariff – January 1997

Service – Water Charges	New Charges (K)
Metered:	
Domestic user – medium- or high-cost:	
first 26 units	14,300
next 10 units, per unit	220
additional units	275
stuck meters	22,000
Industrial/commercial:	
,first 80 units	27,500
next 80 units	385
additional units	440
stuck meters	60,500
Un-metered:	
Domestic users:	
low-cost users per month	6,600
un-graded users per month	2,200
Other charges:	
Water deposit:	
domestic users	27,500
industrial/commercial users	88,000
Reconnection fees:	
domestic users	27,500
industrial/commercial users	88,000
Water connection:	
Domestic users	at cost
Industrial/commercial users	at cost
Illegal water connection:	
Domestic users	220,000
Industrial/commercial users	550,000
Illegal water reconnection:	
Domestic users	220,000
Industrial/commercial users	550,000

Revised Kitwe water tariff – January 1997 (continued)

Service – Sewerage Charges	New Charges (K)
Sewerage charge per point:	
low-cost	6,600
medium or high-cost	10,450
industrial/commercial	18,700
Other charges:	
Sewer unblocking:	
low-cost	3,300
medium or high-cost	5,500
industrial/commercial	13,200
Sewer connection:	
domestic	at cost
industrial/commercial	at cost
Removal of waste:	
domestic per load	19,800
industrial/commercial per load	27,500
outside Kitwe	as advised

Note: The ownership of water meters is vested in the landlords of properties.

Appendix Table 10 New tariffs – Lusaka Water and Sewerage Company Ltd., May 1997

NEW TARIFFS

Effective 1st May, 1997

A. WATER TARIFF - MONTHLY CHARGES

TARIFF 1 COMMERCIAL AND INDUSTRIAL CONSUMERS

	K	N
STANDING CHARGE (For Metered Supplies)	10,000.00	
0 - 100,000 Litres per 1,000 Litres	470.00	
100,001 - 170,000 Litres per 1,000 Litres	750.00	
170,001 Litres and over per 1,000 Litres	1,100.00	

TARIFF 2 HIGH DENSITY RESIDENCES WITH COMMUNAL TAPS

Upgraded Areas	9,000.00

TARIFF 3 HIGH DENSITY RESIDENCES WITH INDIVIDUAL TAPS

Per House / Month	9,000.00

TARIFF 4 LOW DENSITY, COUNCIL RESIDENCES AND SPECIAL CONNECTIONS

STANDING CHARGE (For Metered Supplies)	4,000.00
0 - 6,000 Litres per 1,000 Litres	180.00
6,001 - 66,000 Litres per 1,000 Litres	350.00
66,001 - 100,000 Litres per 1,000 Litres	470.00
100,001 - 170,000 Litres per 1,000 Litres	750.00
170,001 Litres and over per 1,000 Litres	1,100.00

B. WATER CONNECTION CHARGES

All connections to be charged at cost

C. WATER DEPOSITS

(i) Industrial commercial and special connection	260,000.00
(ii) Low and medium Density Residences	50,000.00
(iii) High Density Residences with Individual taps	20,000.00
(iv) High Density Residences with communal taps	20,000.00

D. WATER RECONNECTION FEES

(i) Commercial and Industrial Consumers	190,000.00
(ii) All other areas	50,000.00
(iii) Additional penalty where supply is disconnected due to misuse of water	100,000.00

E. METER INSPECTION FEES

	K	N
Standard fee (to be refunded if the meter is found to be faulty)	40,000.00	

F. SEWERAGE TARIFF: MONTHLY CHARGES

DOMESTIC SEWERAGE TARIFF

0 - 6,000 Litres per 1,000 Litres	180.00
6,001 - 66,000 Litres per 1,000 Litres	350.00
66,001 - 100,000 Litres per 1,000 Litres	470.00
100,001 - 170,000 Litres per 1,000 Litres	750.00
170,001 Litres and over per 1,000 Litres	1,100.00

PREMISES DISCHARGING TRADE EFFLUENT

(Charges per 1,000 Litres of Effluent Discharged)

Tariff C up to 1,000 parts per Million B.O.D.	800.00
Tariff D 1,001 - 2,000 parts per Million B.O.D.	1,200.00
Tariff E 1,001 - 2,000 parts per Million B.O.D.	1,700.00

✻ B.O.D. = Biochemicals Oxygen Demand

G. SEWERAGE DISCHARGE BY VACUUM TANKERS

Per 1,000 Litres	2,500.00

H. SEWERAGE CONNECTION CHARGES

All Sewerage connections to be charged out at cost

I. ILLEGAL CONNECTION CHARGES

WATER

Commercial and Industrial Areas	500,000.00
Residential Areas	100,000.00

SEWER

Commercial and Industrial Areas	500,000.00
Residential Areas	150,000.00

Customers will receive bills based on the new tariffs in June/July 1997

Lusaka Water and Sewerage Company Limited

Index

Printed and bound by CPI Group (UK) Ltd, Croydon, CR0 4YY

22/10/2024

01777640-0004